Rotor Pole Pattern Topology Technology for Permanent Magnet Electric Machine

Pengjie Xiang · Xinghua He · Liang Yan

Rotor Pole Pattern Topology Technology for Permanent Magnet Electric Machine

Springer

Pengjie Xiang
Beihang University
Beijing, China

Liang Yan
Beihang University
Beijing, China

Beihang Ningbo Institute of Technology
Ningbo, China

Tianmushan Laboratory
Hangzhou, China

Xinghua He
Beihang University
Beijing, China

Beijing Information Science
and Technology University
Beijing, China

ISBN 978-981-95-2017-6 ISBN 978-981-95-2018-3 (eBook)
https://doi.org/10.1007/978-981-95-2018-3

This Springer imprint is published by the registered company Springer Nature Singapore Pte Ltd.
The registered company address is: 152 Beach Road, #21-01/04 Gateway East, Singapore 189721, Singapore

The book is dedicated to authors' families

Foreword

In recent years, electrification technologies have become a key enabler for many applications including but not limited to renewable energy generation, robotics, green transportation including road vehicles, aviation, rail and marine. Electrification enables efficient energy generation, conversion, distribution and utilization of energy leading to advantages at application level such as improved sustainability and performance. The electric machine is one of the major components of electrification technology. It converts electrical energy to mechanical and vice versa and has a major impact on the performance of the electrified system in terms of efficiency, weight, volume, cost, embedded carbon and reliability.

This monograph offers one good reference for the structure design and performance improvement of electric machines to engineering students, academic researchers and practicing industrial engineers alike. It provides unique insight into the design methodology of electric machines and presents some innovative, very high-performance electric machines. The book mainly focuses on the reconstruction of topology, magnet shaping, magnetization of materials, such as the magnet irregular shaping method, the combination of magnet and iron poles, and the injection method for specific magnetic harmonic terms, to improve the output performance of electric machines. Some unconventional hybrid structures that are covered in the book aim to enhance the flux-focusing effect, the self-shielding capability, and reduce specific magnetic harmonics or reconstructed flux path. The final target is to increase the torque output, depress the torque ripple and decrease the system mass. These parameters are major concerns of the designers of electric machines. In addition, the proposed multi-material topology optimization method based on variable density representation is quite interesting and very useful, because it does not require strict constraints for the pole shape, the material property, the magnetization directions and the material boundaries. It provides a powerful tool for the global design optimization of electric machines.

There are many books covering the operating principle, the performance modeling and the motion control of electric machines. However, this book systematically covers the pole pattern technology and presents various topologies and design methods to improve the machine performance. Actually, in the past years, the research group has proposed various innovative heterogeneous and irregular pole patterns to achieve impressive performance improvement of electric machines.

This monograph is particularly valuable as a reference for electrical machines design engineers and researchers in that it is complementary to previously published papers and books, and it covers analytical, numerical and experimental methodologies of various high-performance pole pattern topologies, and the design optimization method as well. The examples given in the book and the associated design methodologies presented will surely spawn further creative designs of electric machines for different applications.

Nottingham, UK
December 2024

Christopher Gerada

Preface

Permanent-magnet (PM) brushless machines are widely used in many industrial applications, such as electric propulsion, traction systems and robotics, due to their high torque density, high working efficiency, high reliability, and low noise. Increasing torque, reducing torque ripple, and achieving a lightweight structure are the key issues for the development of PM brushless machines. Typically, there are two ways to improve the torque output of PM machines: increasing current input and increasing magnetic flux density in the air gap. Although increasing current input is a straightforward way to improve torque output, it may lead to high temperatures and cause heat dissipation problems. Additionally, employing PM with large flux-emitting area also helps to increase flux density and force output. However, this approach may increase the system volume further. This monograph aims to increase magnetic flux density and torque output by reorganizing the rotor magnet poles. This book is dedicated to the rotor pole pattern topology technology for PM machines for the first time and introduces different pole patterns according to their functions and application scenarios. These magnet patterns can not only effectively reduce torque ripple and improve torque, but also increase the utilization of magnets, thereby improving the efficiency of permanent-magnet machines.

Chapter 1 introduces the fundamental concepts of PM machines and lists various typical magnetic pole patterns along with their typical applications. According to the type of magnetic pole configuration, they are divided into three types: radial pole pattern, spoke-type pole pattern, Halbach array and their variants.

Chapter 2 introduces two methods for effectively reducing torque ripple. A novel quasi-cylindrical pole pattern, a radial pole pattern, has been introduced as a promising design to reduce torque ripple. The eccentric distance of one quasi-cylindrical is maximized to reduce high-order magnetic harmonics and hence achieve lower cogging torque and torque ripple. Additionally, the fundamental flux density in air gap can be enhanced further. The semi-embedded magnet poles in the rotor can eliminate rotor sleeve, which is commonly used in conventional surface-mounted PM (SPM) machines, thereby reducing the airgap reluctance and further boosting torque output. The quasi-cylindrical magnets also offer other advantages, such as reduced rotor eddy current losses and increased resistance to demagnetization, compared with traditional radial and eccentric pole configurations. Spoke-type PM machine is generally considered as a promising candidate to achieve strong flux-focusing capability and additional reluctance torque. However, large torque ripple can also be founded in this spoke-type design, due to their abundant high-order magnetic harmonics. To achieve low torque ripple of the spoke-type PM machine, a hybrid irregular magnetic pole pattern, that combines quasi-spoke magnets with surface-mounted magnets, is introduced. This pole pattern not only reduces torque ripple but also maintains high torque output of spoke-type design. The quasi-spoke-type magnets, along with surface-mounted PMs featuring multi-directional magnetization, significantly enhance average torque. Furthermore, surface-mounted magnets with parallel magnetization are instrumental in diminishing high-order air-gap magnetic harmonics, leading to a smooth torque output. Following that, the injecting model of third-order harmonics is presented in this chapter, which can be employed in SPM machine to reduced torque ripple while maintaining relatively high torque output. The study analyzes the effects on the magnetic field, electromagnetic force, and vibration in an integral-slot surface-mounted PM machine. It starts by establishing the shape functions for these harmonics and then assesses their impact on air-gap flux density and torque. The research further examines the radial electromagnetic force using 2D-fast Fourier transformation (FFT) to identify the most influential components and how they are affected by different PM edge thicknesses. Finally, a vibration synthesis model predicts the frequency response, showing that the optimal PM edge thickness can reduce both low- and high-frequency vibrations, thereby enhancing machine performance.

Chapter 3 introduces a concentrated-flux-type PM machine with irregular magnets and iron poles. This pole pattern leverages the flux-focusing effect of quasi-trapezoid magnets with opposing magnetization directions and the self-shielding capabilities of quasi-triangular iron poles, which helps to eliminate rotor yoke and reduce rotor mass. Drawing on the principles of linear machines, the chapter also discusses the potential of a novel compound Halbach array. This compound array, consisting of an outer quasi-Halbach layer and an inner axially magnetized pattern, is effective in reducing rotor mass and improving output performance by enhancing self-shielding and eliminating the need for high-density back iron, thus contributing to low mover mass and good dynamic performance.

Chapter 4 introduces a multi-material topology optimization method for SPM machine. And the magnet profile of rotor pole pattern can be automatically obtained according to different objective functions employed this method. Compared with conventional topology optimization method, the proposed multi-material method improves computational efficiency by refining the objective function and clarifying volume constraints, through an integrated approach of size and topology optimization. The design method is validated through theoretical, numerical, and experimental studies. It is implemented in a surface-mounted rotor design to validate this method. And a research prototype is developed, and experiments are conducted to confirm the effectiveness of the optimization method. Additionally, a fixed-output torque optimization method is proposed to simplify the calculation process and achieve high precision, particularly for stator design under rated conditions. Sensitivity analysis on the stator teeth width reveals its impact on motor performance, leading to optimal solutions for both torque and efficiency. The efficiency-optimized solution proves superior, indicating that motor performance is not directly tied to the B-H knee point of stator core. The method presented in this chapter can significantly enhance motor performance, offering a superior alternative to traditional design approaches.

Chapter 5 concludes the major works in this monograph. More subsequent works could be conducted on the proposed novel designs of PM machines. The same design methods and analysis approaches could be implemented in other linear and rotary machines.

Beijing, China
December 2024

Liang Yan

Acknowledgements This monograph was funded by the National Natural Science Foundation of China (NSFC) under grant 52130505, U2570232, the Zhejiang Provincial Natural Science Foundation of China under Grant No. LD24E050005, Ningbo key scientific and technological project under grant 2022Z040, Fundamental Research Funds for the Central Universities, and Science and Technology on Aircraft Control Laboratory.

Contents

Acronyms

AFM	Axial flux machine
CM	Coenergy method
CPM	Consequent-pole PM
CPMIP	The novel concentrated-flux-type PM machine with irregular magnets and iron poles
EMF	Electromotive force
ERRFM	External rotor radial flux machine
FFT	Fast Fourier transformation
GCMMA	Globally convergent method of moving asymptotes
IPM	Interior PM
IRRFM	Inner rotor radial flux machine
MMF	Magnetomotive force
OMD	Optimized material distribution
PM	Permanent magnet
PMSM	permanent magnet synchronous motor
QCPP	Quasi-cylindrical pole pattern
RFM	Radial flux machine
SPM	Surfacc-mounted PM
TFM	Transverse flux machine
THD	Total harmonic distortion
UMF	Unbalanced magnetic force

List of Figures

List of Tables

Chapter 1
Introduction

Abstract Permanent-magnet (PM) brushless synchronous machines have been widely used in many industrial applications, such as electric propulsion and traction systems, due to their high working efficiency, high reliability and low noise. Increasing torque, reducing torque ripple, and lightweight structure are the key issues for the development of PM machines. The rotor pole pattern is one of the important aspects that influence the torque output significantly. In this chapter, the advantages and applications of PM machines with typical pole patterns are presented and discussed. Then, the main objective of this monograph is presented.

1.1 Typical Magnet Pole Pattern Topology

Magnet pole pattern refers to the arrangement of rotor magnets with different shapes, thicknesses and magnetization directions, which exerts significant influence on the airgap flux density and hence the output performance of PM blushless machines such as average torque, torque ripple and unbalanced pull force [1–4]. The magnet pole patterns are generally classified into surface-mounted and interior pole patterns. The magnet patterns may vary a bit for different rotor shapes such as the cylindrical shape of radial-flux PM machines, and the disc shape of axial-flux PM machines, but the fundamental concepts and characteristics are similar. Therefore, for simplification, the subsequent discussion of magnet pole patterns will mainly focus on the cylindrical shape of RFM. The results can be extended to other rotor shapes with similar pole patterns.

1.1.1 Radial Pole Pattern and Its Variants

The pole pattern of fan-shaped magnet with radial magnetization or radial pole pattern in Fig. 1.1a is the most popular pole patterns used in electric PM machines due to its easy design and simple fabrication [5–7]. Some researches of radial pole

P. Xiang et al., *Rotor Pole Pattern Topology Technology for Permanent Magnet Electric Machine*, https://doi.org/10.1007/978-981-95-2018-3_1

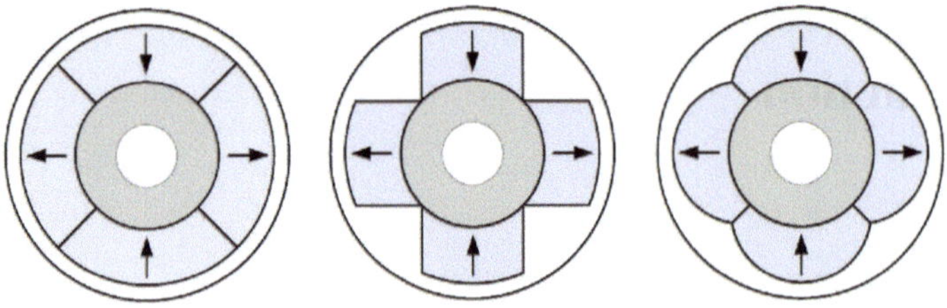

(a) Radial pole (b) Parallel sided pole (c) Sine-pole

Fig. 1.1 Radial pole pattern and its variants [8]

pattern focus on the magnet shaping methods to modify airgap flux density distribution and hence achieve high torque output or low torque ripple. For example, the fan-shaped magnet can be cut into parallel sided arc-shaped magnets as shown in Fig. 1.1b [8] or the sinusoidal shaped magnets as shown in Fig. 1.1c [9] to reduce the torque ripple, because the radii of inner and outer magnet surfaces can be optimized with segmented sections to achieve sinusoidal flux density distribution. However, when the fan-shaped magnet is cut in these two types, the method by using feedback function to introduce all orders of harmonics into the magnet shape. The results show that the fundamental, the third and the fifth harmonics make great contributions to improve torque quality, while the seventh harmonic and higher orders are insignificant for torque improvement [10]. For a more accurate harmonics factor, Hu et al. improved the harmonics injection method by taking the nonideal factors into consideration such as the stator slotting, the pole leakage, the pole edge effect and so on [11]. Zeng et al. injected the sine, inverse cosine and tangent harmonic compensations into magnet shaped to avoid significant drop of the average torque when reducing the torque ripple [12]. The magnet shapes of the three injection methods are shown in Fig. 1.2. The comparison between the conventional radial pole pattern and the three injection magnet pole patterns shows that the torque ripple of the sine shaped and the inverse-cosine shaped machines are reduced by 14.5–15% compared with electric machine with radial pole pattern, while their average torques have a loss of 4% under a small PM volume.

(a) Sine harm (b) Inverse cosine harm (c) Tangent harm

Fig. 1.2 Magnet pole shapes with different harmonic injection method [12]

1.1.2 Spoke-Type Pole Pattern and Its Variants

Figure 1.3a shows one typical spoke-type pole pattern consisting of rectangular magnets with tangential magnetization and quasi-trapezoidal iron block [13]. The magnetic flux of each rotor pole is generated by two magnets with large cross section to enhance airgap flux density. In other words, the spoke-type pole pattern generally has very strong electric flux-focusing effect. Besides, the employment of interpolar iron block with high permeability could result in large difference between the d- and q-axis inductances, which helps to improve the reluctance torque [14]. Therefore, the average torque of the electric PM machines with the spoke-type pole pattern is generally high due to the electric flux-focusing effect and the reluctance torque. In addition, the enhanced d- axis inductance helps to obtain great flux-weakening performance and hence large constant speed range.

The recent studies of the spoke-type pole pattern main focus on the shape modification of magnets and iron blocks to achieve high average torque, low torque ripple, large constant power speed range, and low flux leakage etc. For example, Aydin et al. proposed a spoke-type pole pattern with sinusoidal rotor segments, and it was employed in a coreless dual-rotor AFM.145 The sinusoidal rotor segment is actually an eccentrical iron block as shown in Fig. 1.3a, which leads to variable airgap thickness. Thus, sinusoidal and high airgap magnetic field can be obtained by adjusting the structural parameters such as segments/pole pitch ratio and airgap depth. Ge et al. proposed an alternate airspace barrier for the spoke-type pole pattern with the eccentrical iron blocks as shown in Fig. 1.3b. The unipolar flux leakage can be reduced by

Fig. 1.3 Spoke-type pole pattern and its variants [13, 15, 16, 18, 19]

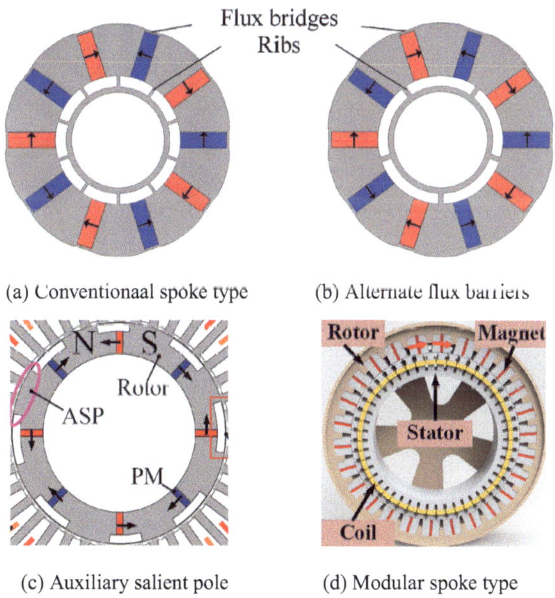

the alternate airspace barrier, and the airgap magnetic resistance can be decreased by the eccentrical iron blocks. The average torque of the spoke-type PM machines can thus be further increased [15, 16]. For the optimization of the spoke-type pole pattern with eccentrical iron blocks, Kong et al. developed a subdomain analytical method [17]. The eccentric iron blocks were approximated by multiple layers of arc-shaped subdomains in the analytical method to predict the torque output. Chen et al. proposed a similar shaped iron block called auxiliary salient poles (ASPs) as shown in Fig. 1.3c [18]. The difference between the eccentric iron and the auxiliary salient iron is that the former protrudes toward the airgap while the latter is recessed. From the perspective of the electromagnetic field, they have the same functions of the modification of the airgap flux density distribution. The torque ripple can be suppressed by the unequal and shifted ASPs. Zhang et al. proposed a modular spoke-type pole pattern with rectangular iron shown in Fig. 1.3d [19]. The tangential magnetized magnets are separated with two iron blocks to form a sandwiched structure. The sandwiched structures are isolated with nonmagnetic supporting cells. Thus, one piece of magnet can produce one pair of magnet poles, which is entirely different from the conventional spoke-type machines. The results show that the large number of slot-pole numbers, the high torque output and strong flux weakening capability can be achieved, compared with conventional spoke-type machines [20, 21].

Another method to improve the output performance of the spoke-type PM machines is to modify their magnet shape. For example, Xia et al. proposed a two-layer spoke-type machine as shown in Fig. 1.4a. The single rectangular magnet in the conventional spoke-type pole pattern is divided into three layers. It can not only

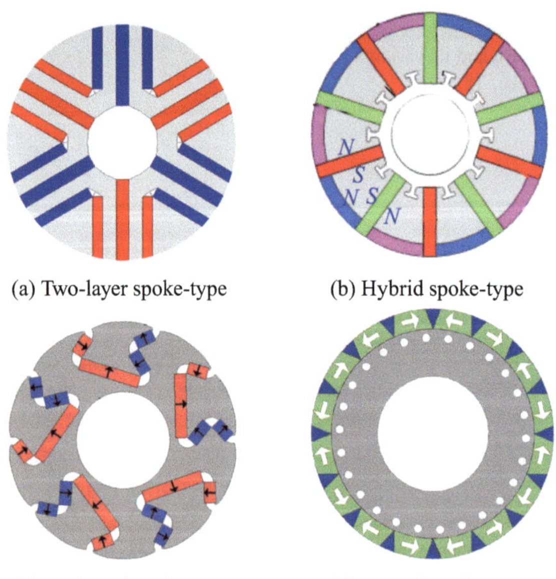

(a) Two-layer spoke-type (b) Hybrid spoke-type

(c) U-shaped spoke-type (d) Trapzd. spoke-type

Fig. 1.4 Spoke-type pole pattern and its variants [22, 23, 25, 26] increase the airgap flux density and the average torque, but also achieve more sinusoidal airgap magnetic field and thus lower torque ripple compared with conventional spoke-type pole patterns [22]. The influence of the size and the location of magnets was also investigated in the paper. In addition, Liu et al. proposed hybrid spoke-type pole pattern shown in Fig. 1.4b, and the performance comparison between the hybrid pole pattern and the conventional spoke-pole pattern were conducted. The addition of the surface-mount magnets helps to obtain higher torque density and stronger magnet anti-demagnetization performance than the conventional spoke-type magnet patterns [23]. Chai et al. proposed a spoke-type pole pattern with trapezoid magnets along with an improved subdomain analytical method [24]. The electromagnetic performance of the electric PM machines can be effectively improved due to the large magnetic flux generated by the trapezoidal magnet with large cross section. Besides, the asymmetric U-shaped spoke-type pole pattern shown in Fig. 1.4c is another choice for torque improvement [25]. The magnetic flux of each pole in the U-shaped spoke-type rotor is produced by four magnets to achieve strong electric electric flux-focusing effect. Moreover, the magnets are locked inside the main rotor core to remove the iron bridge in the conventional spoke-type rotors. Thus, the flux leakage can be reduced and high torque output can be achieved. Yan et al. proposed a spoke-type rotor with special quasi-trapezoidal magnets and quasi-triangular irons, as shown in Fig. 1.4d [26]. The research results show that it can not only enhance the average torque, but also reduce the rotor mass, because of the electric electric flux-focusing effect and the self-shielding effect.

Overall, the spoke-type pole pattern helps to achieve high torque output due to its strong electric electric flux-focusing effect and high reluctance torque. The reduction of flux-leakage, torque ripple and rotor mass is the main challenge for the design of the spoke-type pole patterns.

1.1.3 Halbach Array and Its Variants

Halbach array and its variants Fig. 1.5a shows the most typical Halbach array consisting of radially and tangentially magnetized magnets simultaneously. Each pole of the Halbach array is composed of two magnet segments, and thus it is called a two-segment Halbach array [27]. Compared with the radial pole pattern, the two segmented Halbach array has strong electric electric flux-focusing effect and self-shielding effect. The former helps to enhance the airgap flux density and obtain high torque output, and the latter contributes to weakening the magnetic field in the rotor yoke and achieving thin rotor yoke and low rotor mass [28]. Therefore, many studies focus on the improvement of the electric electric flux-focusing effect and self-shielding effects. There are two typical approaches, i.e., muti-segment Halbach array and irregular shaped Halbach array. For example, the tangentially magnetized magnets in the two-segment Halbach array can be further divided into two magnet segments with oblique magnetization, forming three segmented Halbach array as shown in Fig. 1.5b. The oblique magnetization makes the airgap magnetic field

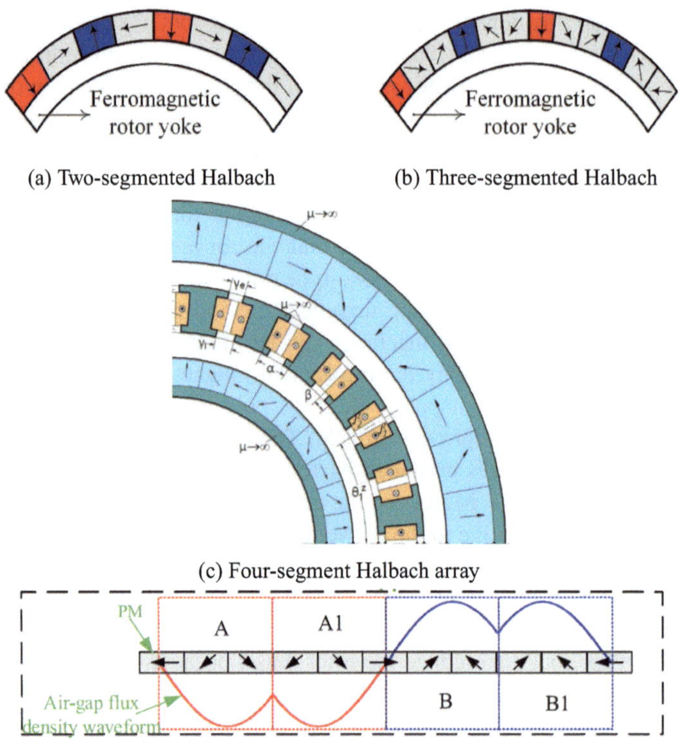

(a) Two-segmented Halbach (b) Three-segmented Halbach

(c) Four-segment Halbach array

(d) Five-segmented Halbach array

Fig. 1.5 Halbach array and its variants [30, 32]

more sinusoidal and hence achieves smooth torque output. Besides, the magnetic flux of each pole of the three segmented Halbach array is generated by three magnets, which results in a higher average torque compared with the two segmented Halbach. Zhu et al. proposed a three-segment Halbach array with unequal side-magnet height, and the results show that the pole pattern can achieve low torque ripple, relatively large torque and high magnet usage efficiency simultaneously [29]. In addition, a dual-rotor RFM with four-segment Halbach array in Fig. 1.5c [30] and a dual-rotor AFM with four-segment Halbach array [31] are developed, respectively, to improve the system torque density. Zhang et al. proposed a five-segmented Halbach array with special magnetization shown in Fig. 1.5d for a five-phase PM machine. The third harmonics can be injected into the airgap magnetic field to increase the system torque density [32]. The increase of the number of magnet segments generally helps to improve the electric electric flux-focusing effect and reduce the flux leakage of the rotor yoke [27, 33]. However, the fabrication cost of these muti-segment Halbach array will increase significantly, and the magnet assembly of the array is also challenging.

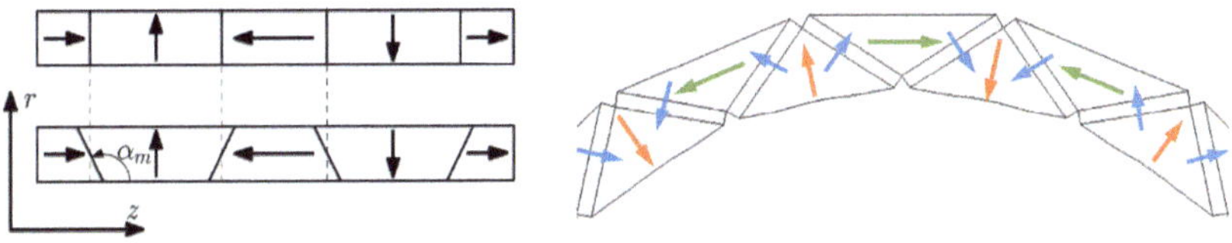

(a) Trapezoidal-shaped (b) Blending-shaped

Fig. 1.6 Halbach arrays with irregular magnets [35, 36]

Alternatively, the Halbach array with irregular shaped magnets is an alternative option to improve the flux-focusing and self-shielding effects. Potenziani et al. proposed a two-segment Halbach array with triangular magnets to achieve very little flux leakage in the rotor yoke and low rotor mass [34]. Meessen et al. proposed a trapezoidal-shaped two-segment Halbach array shown in Fig. 1.6a [35], and a superposition analytical method was developed to predict the airgap flux density distribution. The results show that the sinusoidal airgap magnetic field and the smooth torque output were achieved in the electric PM machine. Dong et al. proposed a blending-shaped Halbach array shown in Fig. 1.6b, and the output performance of the electric machine was compared with the two-segment and three-segment Halbach arrays [36]. The comparison shows that the blending-shaped Halbach array has the highest fundamental airgap flux density and highest average torque.

Overall, Halbach array is a promising choice for high torque density PM machines due to the average torque improvement by the electric electric flux-focusing effect and the rotor mass reduction by the self-shielding effect.

1.2 Some Applications of Typical Pole Patterns

1.2.1 Robot Joint

Electric machines have been become the mainstream for robot joints, e.g., the Optimus, a humanoid robot introduced by Tesla, is expected to have 40 electromechanical actuators of which 12 are in the arms, 2 each in the neck and torso, 12 in the legs, and 12 in the hands [40]. High torque density, high working efficiency and low inertia of joint PM machines are the main requirements for these robot applications, such as legged robots, collaborative robots, humanoid robots etc., to obtain high dynamics and low total cost of transport [37, 38]. High-speed inner-rotor radial-flux PM machines with high-power density are the most popular choice for robot joint modules mainly consisting of High-speed machine, harmonic reducer, integrated controller and encoder, as shown in Fig. 1.7 [39, 41]. To further improve torque density and a more compact structure, external-rotor radial-flux PM machines with large inner hollow space are generally utilized. For example, as shown in Fig. 1.8, Xiaomi

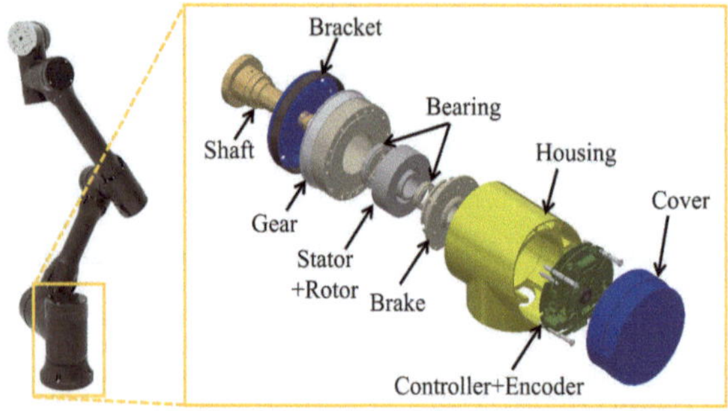

Fig. 1.7 Cooperative robot system and its actuator [41]

Fig. 1.8 CyberGear, a robot joint module, developed by Xiaomi Corporation [47]

Corporation developed a robot joint module, CyberGear, mainly consisting of one ERRFM and the integrated reduce [47]. The system volume can be reduced well, and the peak torque density of 37.85 Nm/kg is achieved. The similar structure can be found in [44]. Additionally, in [45, 46], an axial-flux machine with double-side asynchronous rotor is developed for legged robots, as shown in Fig. 1.9. To drive the lower leg and thigh of the dog robot simultaneously, the two different slot-pole configurations are adopted on two sides of the axial-flux machine. Compared with conventional joint module in Fig. 1.9a, two radial-flux machine connected in series, the joint volume of the dog robot can be reduced greatly with the double-side asynchronous rotor.

The bread-loaf and eccentric pole patterns, having more sinusoidal air-gap magnetic field, can be widely found in these radial-flux joint modules used in various robots, such as the Cheetah robot [37, 39], wearable robot [42] and articulated robot [43], to achieve smooth torque, low core loss and high-precision controlling performance.

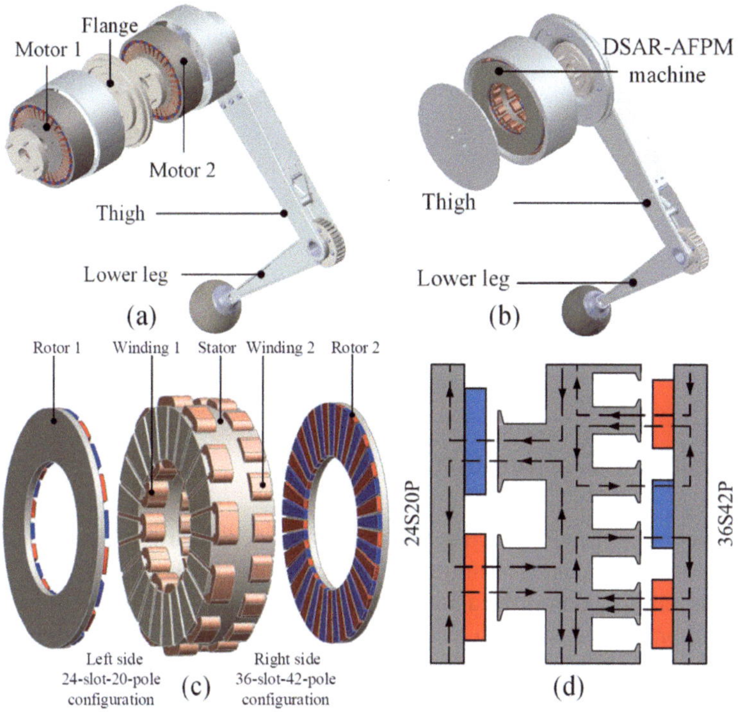

Fig. 1.9 Structure of the leg in a robot dog and the axial-flux machine with the double-side asynchronous rotor [45, 46]. **a** Traditional leg structure. **b** Novel leg structure. **c** Axial-flux machine with double-side asynchronous rotor. **d** Magnetic flux in the axial-flux machine

1.2.2 Aerospace Engineering

IRRFM is an alternative choice for the aviation applications. For example, a slotless IRRFRM with three-segment Halbach array was developed for a drone system that requires four electrical motors of 0.5 kW each for 2 kW propulsion power. The power density of 1.02 kW/kg and the torque density of 1.94 Nm/kg can be reached. Compared with slotted counterparts, the power density of the slotless IRRFM is improved by 52% [48]. Zhang et al. developed one IRRFM for the solar-powered aircraft, and one framed rotor was used to achieve a lightweight structure. Besides, miniature heat pipes are inserted into the stator slots to cool the windings and hence achieve high current density and torque output [49]. NASA designed its first all-electric plane in Fig. 1.10. The distributed electric propulsion system consists of 12 high lift IRRFMs. The Halbach array with high remanence of N48H and the stator and rotor core of Hiperco-50 with high magnetic saturation and low loss were used for the high specific power of 5.64 kW/kg and the high torque density of 6.64 Nm/kg [50, 51]. In recent years, axial flux motors have been developed for the electric propulsion due to their high torque density and compact axial structure. Specifically, the dual-rotor

Fig. 1.10 NASA X57 all-electric plane with 12 IRRFMs [50, 51]

AFMs of EMRAX 268 and YASA 750R were widely employed in various electric aircraft. The torque density is as high as around 11 Nm/kg [52, 53].

1.2.3 Electric Traction Systems

For the traction applications such as electric bicycles and electric vehicles, the electric motors are usually mounted on the rear wheel or the vehicle chassis to provide traction power. The electric motors need not only to be small enough to fit into the compact wheel hub or chassis area, but also to supply sufficient torque output and power to provide strong hill-climbing capability. Thus, high torque and power density are desired for these traction applications. Besides, Fatemi et al. developed a spoke-type IRRFM for the application of Formula-E racing car [57]. The Pareto-optimal method was used to perform geometric optimization of the spoke-type IRRFM to obtain high torque density and high drive-cycle efficiency simultaneously. Besides, Yang et al. developed an in-wheel ERRFM with V-shaped pole pattern, targeting at the e-bike application. The strong electric electric flux-focusing effect and the rotor mechanical construction of the V-shaped pole pattern were utilized to achieve high system torque and power density. Similarly, the V-shaped pole pattern were also used in a dual-rotor RFM implemented into a low-speed direct-drive wheel-hub drive system [58].

A dual rotor RFM with Halbach array was designed to increase the torque density for the application of electrical vehicle [59]. The performance comparison between the Halbach array and the radial pole pattern indicated that an improvement of 16% is observed in the proposed electrical PM motor. Similarly, the dual-rotor RFM with Halbach array can be found in the application of electric super-bike as shown in Fig. 1.11 [54]. The optimal slot-pole ratio of the electric machine was studied in the paper. In addition, because of the spatially independent magnetic and electrical loading, a dual rotor TFM was employed in a hybrid electric vehicle. The dual rotor TFM can provide the speed difference between the internal combustion engine and the reduction gear. Thus, the internal combustion engine can work in the optimum speed torque region independent of the vehicle condition [61].

Fig. 1.11 Electric superbike powered by a dual-rotor RFM [54]

Fig. 1.12 In-wheel traction application powered by YASA [55]

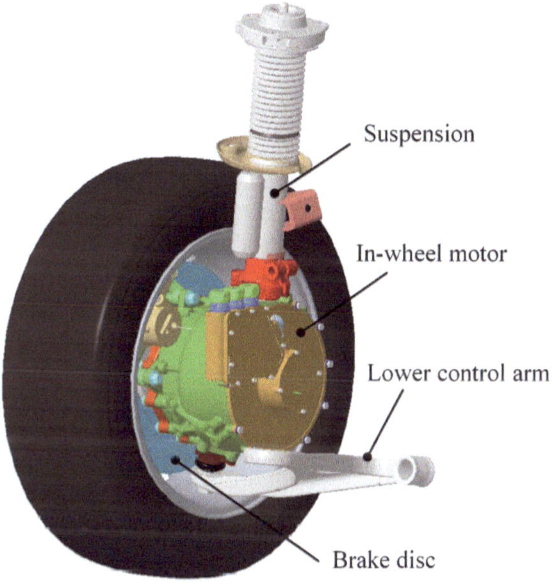

Besides, the axial flux PM motor is also one desirable option for traction systems. For example, Chang et al. developed one yokeless and segmented armature axial flux motor for the in-wheel traction application shown in Fig. 1.12. A U-shaped water-cooling pipe was proposed for the electrical motor to improve the output power from 50 to 65 kW [55]. The single side and other multi-stack AFMs in Sect. 2.2 can be also used in solar cars in Fig. 1.13 to achieve high torque and compact axial space [56].

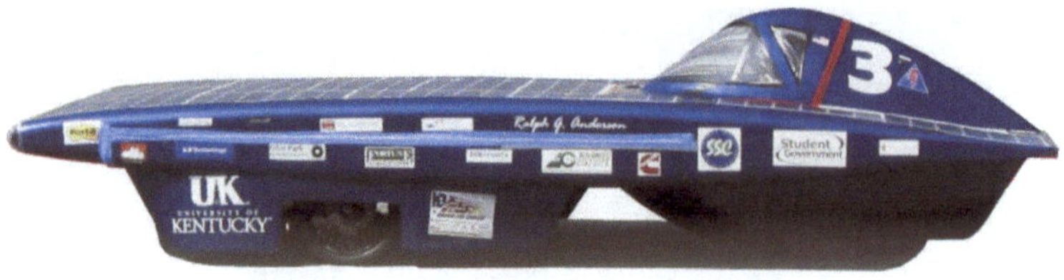

Fig. 1.13 Gato del Sol IV solar car from University of Kentucky [56]

1.3 Objective of the Study

Large torque ripple of PM machines is always undesirable especially for those used in high-performance servo applications. However, some conventional approaches adopted for torque ripple suppression generally result in average torque drop, which may influence the loading capability of PM machines. In addition, as precious and cost resources, improving the utilization rate of rare-earth materials has been a topic of great research for scholars. Meanwhile, improving the utilization rate of materials may further enhance the output characteristics and efficiency of PM machines. Therefore, the objective of this monograph is to introduce several methods for reducing torque ripple, increasing torque, reducing usage of rare-earth magnets, and improving the utilization of the PMs. The remaining chapters of the monograph are organized as follows: Chap. 2 introduces a novel radial-flux surface-mounted PM machine with quasi-cylindrical pole pattern and the hybrid irregular magnetic pole pattern composed of quasi spoke magnets and surface mounted magnets for effectively reducing torque ripple. Chapter 3 introduces two methods for improving output torque while reducing rotor mass. Chapter 4 introduces a multi-material topology optimization method to effectively reduce rotor mass and permanent magnet utilization.

References

1. M. Zhu, W. Hu and N. C. Kar, "Torque-Ripple-Based Interior Permanent-Magnet Synchronous Machine Rotor Demagnetization Fault Detection and Current Regulation," IEEE Transactions on Industry Applications, vol. 53, no. 3, pp. 2795–2804, May-June 2017.
2. Z. Dong, L. Yan, H. Su and I. -M. Chen, "Design and Analysis of a Novel Electromagnetic-Propulsion Rotary Machine With External Blending-Shaped Tessellation Magnet Pattern," IEEE Transactions on Industrial Electronics, vol. 71, no. 2, pp. 1355–1364, 2024.
3. Y. Chen, W. N. Fu, S. L. Ho and H. Liu, "A Quantitative Comparison Analysis of Radial-Flux, Transverse-Flux, and Axial-Flux Magnetic Gears," IEEE Transactions on Magnetics, vol. 50, no. 11, pp. 1–4, Nov. 2014.
4. R. Tsunata, M. Takemoto, J. Imai, T. Saito and T. Ueno, "Comparison of Thermal Characteristics in Various Aspect Ratios for Radial-Flux and Axial-Flux Permanent Magnet Machines," IEEE Transactions on Industry Applications, vol. 59, no. 3, pp. 3353–3367, May-June 2023.
5. R. B. Tavares, E. C. Goltz, P. R. Eckert and L. A. Pereira, "Torque Assessment of Multiphase AFPM Machines With Experimental Validation for YASA Topology," IEEE Transactions on Energy Conversion, vol. 38, no. 1, pp. 111–121, March 2023.

6. J. H. J. Potgieter and M. J. Kamper, "Double PM-Rotor, Toothed, Toroidal-Winding Wind Generator: A Comparison With Conventional Winding Direct-Drive PM Wind Generators Over a Wide Power Range," IEEE Transactions on Industry Applications, vol. 52, no. 4, pp. 2881–2891, July-Aug. 2016.
7. A. Hemeida et al., "A Simple and Efficient Quasi-3D Magnetic Equivalent Circuit for Surface Axial Flux Permanent Magnet Synchronous Machines," IEEE Transactions on Industrial Electronics, vol. 66, no. 11, pp. 8318–8333, Nov. 2019.
8. L. Wu and Z. -Q. Zhu, "Analytical Modeling of Surface-Mounted PM Machines Accounting for Magnet Shaping and Varied Magnet Property Distribution," IEEE Transactions on Magnetics, vol. 50, no. 7, pp. 1–11, July 2014.
9. Yong Li, Jibin Zou and Yongping Lu, "Optimum design of magnet shape in permanent-magnet synchronous motors," IEEE Transactions on Magnetics, vol. 39, no. 6, pp. 3523–3526, Nov. 2003.
10. G. Liu, Y. Zeng, W. Zhao and J. Ji, "Permanent Magnet Shape Using Analytical Feedback Function for Torque Improvement," IEEE Transactions on Industrial Electronics, vol. 65, no. 6, pp. 4619–4630, June 2018.
11. P. Hu et al., "The Modified Model of Third-Harmonic Shaping for a Surface-Mounted Permanent-Magnet Synchronous Motor Under Parallel Magnetization," IEEE Transactions on Industry Applications, vol. 56, no. 5, pp. 4847–4856, Sept.-Oct. 2020.
12. Y. Zeng, M. Cheng, G. Liu and W. Zhao, "Effects of Magnet Shape on Torque Capability of Surface-Mounted Permanent Magnet Machine for Servo Applications," IEEE Transactions on Industrial Electronics, vol. 67, no. 4, pp. 2977–2990, April 2020.
13. Y. Du et al., "Magnetic Field Prediction in Cubic Spoke-Type Permanent-Magnet Machine Considering Magnetic Saturation," IEEE Transactions on Industrial Electronics, vol. 71, no. 3, pp. 2208–2219, March 2024.
14. H. -C. Liu, I. -G. Kim, Y. J. Oh, J. Lee and S. -C. Go, "Design of Permanent Magnet-Assisted Synchronous Reluctance Motor for Maximized Back-EMF and Torque Ripple Reduction," IEEE Transactions on Magnetics, vol. 53, no. 6, pp. 1–4, June 2017.
15. X. Ge, Z. Q. Zhu, J. Li and J. Chen, "A Spoke-Type IPM Machine With Novel Alternate Airspace Barriers and Reduction of Unipolar Leakage Flux by Step-Staggered Rotor," IEEE Transactions on Industry Applications, vol. 52, no. 6, pp. 4789–4797, Nov.-Dec. 2016.
16. N. Elloumi, M. Bortolozzi, A. Masmoudi, M. Mezzarobba, M. Olivo and A. Tessarolo, "Numerical and Analytical Approaches to the Modeling of a Spoke Type IPM Machine With Enhanced Flux Weakening Capability," IEEE Transactions on Industry Applications, vol. 55, no. 5, pp. 4702–4714, Sept.-Oct. 2019.
17. X. Kong, Y. Hua, Z. Zhang, C. Wang and Y. Liu, "Analytical Modeling of High-Torque-Density Spoke-Type Permanent Magnet In-Wheel Motor Accounting for Rotor Slot and Eccentric Magnetic Pole," IEEE Transactions on Transportation Electrification, vol. 7, no. 4, pp. 2683–2693, Dec. 2021.
18. Q. Chen, G. Xu, F. Zhai and G. Liu, "A Novel Spoke-Type PM Motor With Auxiliary Salient Poles for Low Torque Pulsation," IEEE Transactions on Industrial Electronics, vol. 67, no. 6, pp. 4762–4773, June 2020.
19. H. Zhang, W. Hua and Z. Wu, "Modular Spoke-Type Permanent-Magnet Machine for In-Wheel Traction Applications," IEEE Transactions on Industrial Electronics, vol. 65, no. 10, pp. 7648–7659, Oct. 2018.
20. H. Zhang, W. Hua, D. Gerada, C. Gerada, Y. Li and G. Zhang, "Comparative Study on Two Modular Spoke-Type PM Machines for In-Wheel Traction Applications," IEEE Transactions on Energy Conversion, vol. 34, no. 4, pp. 2137–2147, Dec. 2019.
21. H. Zhang, W. Hua, D. Gerada, Z. Xu and C. Gerada, "Analytical Model of Modular Spoke-Type Permanent Magnet Machines for In-Wheel Traction Applications," IEEE Transactions on Energy Conversion, vol. 35, no. 3, pp. 1289–1300, Sept. 2020.
22. B. Xia and P. C. -K. Luk, "Analytical model of open-circuit characteristics of two-layer spoke-type ferrite interior permanent magnet machines," 2017 IEEE International Electric Machines and Drives Conference (IEMDC), Miami, FL, USA, 2017, pp. 1–8.

23. Y. Liu, L. Li, Q. Gao, J. Cao and Z. Sun, "An Analytical Model and Optimization of a Novel Hybrid Rotor Machine for High Torque Density," IEEE Transactions on Energy Conversion, vol. 36, no. 1, pp. 230–241, March 2021.
24. P. Liang, F. Chai, Y. Yu and L. Chen, "Analytical Model of a Spoke-Type Permanent Magnet Synchronous In-Wheel Motor With Trapezoid Magnet Accounting for Tooth Saturation," IEEE Transactions on Industrial Electronics, vol. 66, no. 2, pp. 1162–1171, Feb. 2019.
25. J. Li and K. Wang, "A Novel Spoke-Type PM Machine Employing Asymmetric Modular Consequent-Pole Rotor," IEEE/ASME Transactions on Mechatronics, vol. 24, no. 5, pp. 2182–2192, Oct. 2019.
26. P. Xiang, L. Yan, Y. Guo, X. He, C. Gerada and I. -M. Chen, "A Concentrated-Flux-Type PM Machine With Irregular Magnets and Iron Poles," IEEE/ASME Transactions on Mechatronics, vol. 29, no. 1, pp. 691–702, Feb. 2024.
27. Z. Q. Zhu, Z. P. Xia and D. Howe, "Comparison of Halbach magnetized brushless machines based on discrete magnet segments or a single ring magnet," IEEE Transactions on Magnetics, vol. 38, no. 5, pp. 2997–2999, Sept. 2002.
28. P. Jin, Y. Yuan, Q. Xu, S. Fang, H. Lin and S. L. Ho, "Analysis of Axial-Flux Halbach Permanent-Magnet Machine," IEEE Transactions on Magnetics, vol. 51, no. 11, pp. 1–4, Nov. 2015.
29. Y. Shen and Z. Q. Zhu, "Analysis of Electromagnetic Performance of Halbach PM Brushless Machines Having Mixed Grade and Unequal Height of Magnets," IEEE Transactions on Magnetics, vol. 49, no. 4, pp. 1461–1469, April 2013.
30. D. Golovanov and C. Gerada, "An Analytical Subdomain Model for Dual-Rotor Permanent Magnet Motor With Halbach Array," IEEE Transactions on Magnetics, vol. 55, no. 12, pp. 1–16, Dec. 2019.
31. Z. Zhang, C. Wang and W. Geng, "Design and Optimization of Halbach-Array PM Rotor for High-Speed Axial-Flux Permanent Magnet Machine With Ironless Stator," IEEE Transactions on Industrial Electronics, vol. 67, no. 9, pp. 7269–7279, Sept. 2020.
32. L. F. Zhang, K. Wang, H. Y. Sun and S. S. Zhu, "Multiphase PM Machines With Halbach Array Considering Third Harmonic Flux Density," IEEE Transactions on Industrial Electronics, vol. 66, no. 12, pp. 9184–9193, Dec. 2019.
33. X. Zhang, Z. Yuan, Q. Yang, Y. Li, J. Zhu and Y. Li, "Coil Design and Efficiency Analysis for Dynamic Wireless Charging System for Electric Vehicles," IEEE Transactions on Magnetics, vol. 52, no. 7, pp. 1–4, July 2016.
34. Potenziani II E, Leupold H A, Basarab D J, "A novel self-shielding permanent-magnet rotor assembly," Journal of Applied Physics, vol. 64, no. 10, pp. 5986–5987, 1988.
35. K. J. Meessen, B. L. J. Gysen, J. J. H. Paulides and E. A. Lomonova, "Halbach Permanent Magnet Shape Selection for Slotless Tubular Actuators," IEEE Transactions on Magnetics, vol. 44, no. 11, pp. 4305–4308, Nov. 2008.
36. Z. Dong, L. Yan, H. Su and I. -M. Chen, "Design and Analysis of a Novel Electromagnetic-Propulsion Rotary Machine With External Blending-Shaped Tessellation Magnet Pattern," IEEE Transactions on Industrial Electronics, vol. 71, no. 2, pp. 1355–1364, Feb. 2024.
37. T. -W. Lee, D. -K. Hong and T. -U. Jung, "High-speed, high-power motor design for a four-legged robot actuator optimized using the weighted sum and response surface methods," CES Transactions on Electrical Machines and Systems, vol. 5, no. 3, pp. 224–231, Sept. 2021.
38. S. Seok et al., "Design Principles for Energy-Efficient Legged Locomotion and Implementation on the MIT Cheetah Robot," IEEE/ASME Transactions on Mechatronics, vol. 20, no. 3, pp. 1117–1129, June 2015.
39. D. -K. Hong, W. Hwang, J. -Y. Lee and B. -C. Woo, "Design, Analysis, and Experimental Validation of a Permanent Magnet Synchronous Motor for Articulated Robot Applications," IEEE Transactions on Magnetics, vol. 54, no. 3, pp. 1–4, March 2018.
40. Malik, Ali Ahmad, Tariq Masood, and Alexander Brem, "Intelligent humanoids in manufacturing to address worker shortage and skill gaps: Case of Tesla Optimus." arXiv preprint arXiv:2304.04949, 2023.
41. P. T. Luu, J. -Y. Lee, J. -H. Lee and J. -W. Park, "Electromagnetic and Thermal Analysis of Permanent-Magnet Synchronous Motors for Cooperative Robot Applications," IEEE Transactions on Magnetics, vol. 56, no. 3, pp. 1–4, March 2020.

42. S. -W. Hwang, J. -W. Chin and M. -S. Lim, "Design Process and Verification of SPMSM for a Wearable Robot Considering Thermal Characteristics Through LPTN," IEEE/ASME Transactions on Mechatronics, vol. 26, no. 2, pp. 1033–1042, April 2021.
43. J. Lee et al., "Design of the joint motor for an articulated robot considering joint load characteristics," Energies, vol. 14, no. 20, pp. 6690, 2021.
44. G. Zhang et al., "Design of a high torque density robot joint and analysis of force control method applied for a light exoskeleton.," Electronics, vol. 12, no. 2, pp. 397, 2023.
45. R. Huang, Z. Song, Z. Dong, Y. Liu and C. Liu, "Hybrid Optimization and Comparative Analysis of a Novel Double-Side Asynchronous Rotor Axial-Flux Permanent Magnet Machine," IEEE Transactions on Transportation Electrification, vol. 10, no. 2, pp. 3373–3382, June 2024.
46. R. Huang, Z. Song, Z. Dong, Y. Liu and C. Liu, "Design of A New Double Side Axial-Flux Actuator for Robot Dog," 2022 International Conference on Electrical Machines (ICEM), Valencia, Spain, 2022, pp. 900–906.
47. Corporation, X.: CyberGear motor. https: //www.mi.com/cyber-gear (2024)
48. M. S. Islam, R. Mikail and I. Husain, "Slotless Lightweight Motor for Aerial Applications," IEEE Transactions on Industry Applications, vol. 55, no. 6, pp. 5789–5799, Nov.-Dec. 2019.
49. X. Zhang, L. Li and C. Zhang, "Mass Optimization Method of a Surface-Mounted Permanent Magnet Synchronous Motor Based on a Lightweight Structure," IEEE Access, vol. 8, pp. 40431–40444, 2020.
50. D. Hall et al., "Development of a Maxwell X-57 high lift motor reference design," AIAA Propulsion and Energy, 2019.
51. A. Dubois et al., "Design of an electric propulsion system for SCEPTOR's outboard nacelle," 16th AIAA aviation technology, integration, and operations conference, 2016.
52. Emrax: Yasa. Applications. https://yasa.Com/applications/ (2023)
53. Emrax: Emrax.e-motors:emrax-268. https://emrax.Com/e-motors/emrax-268/ (2023)
54. Y. Song, Z. Zhang et al., "Analysis and reduction of cogging torque in direct-drive external-rotor permanent magnet synchronous motor for belt conveyor application," IET Electric Power Applications, vol. 15, pp. 668–680, 2021.
55. H. Mamur et al., "Design and fabrication of an outer rotor permanent magnet synchronous generator with fractional winding for micro-wind turbines," IET electric power applications, vol. 14, no. 12, pp. 2273–2282, Nov.-Dec. 2020.
56. N. Taran, V. Rallabandi, G. Heins and D. M. Ionel, "Coreless and Conventional Axial Flux Permanent Magnet Motors for Solar Cars," IEEE Transactions on Industry Applications, vol. 54, no. 6, pp. 5907–5917, Nov.-Dec. 2018.
57. A. Fatemi, D. M. Ionel, M. Popescu, Y. C. Chong and N. A. O. Demerdash, "Design Optimization of a High Torque Density Spoke-Type PM Motor for a Formula E Race Drive Cycle," IEEE Transactions on Industry Applications, vol. 54, no. 5, pp. 4343–4354, Sept.-Oct. 2018.
58. Z. Zhang, "A Compact High Torque Density Dual Rotor Permanent Magnet In-Wheel Motor With Toroidal Windings," 2019 22nd International Conference on Electrical Machines and Systems (ICEMS), Harbin, China, 2019.
59. B. V. R. Kumar and K. S. Kumar, "Design of a new Dual Rotor Radial Flux BLDC motor with Halbach array magnets for an electric vehicle," 2016 IEEE International Conference on Power Electronics, Drives and Energy Systems (PEDES), Trivandrum, India, 2016, pp. 1–5.
60. D. Golovanov, A. Galassini, L. Flanagan, D. Gerada, Z. Xu and C. Gerada, "Dual-Rotor Permanent Magnet Motor for Electric Superbike," 2019 IEEE International Electric Machines and Drives Conference (IEMDC), San Diego, CA, USA, 2019, pp. 951–956.
61. M. Wang, P. Zheng, C. Tong, Q. Zhao and G. Qiao, "Research on a Transverse-Flux Brushless Double-Rotor Machine for Hybrid Electric Vehicles," IEEE Transactions on Industrial Electronics, vol. 66, no. 2, pp. 1032–1043, Feb. 2019.
62. J. Chang, Y. Fan, J. Wu and B. Zhu, "A Yokeless and Segmented Armature Axial Flux Machine With Novel Cooling System for In-Wheel Traction Applications," IEEE Transactions on Industrial Electronics, vol. 68, no. 5, pp. 4131–4140, May 2021.

Chapter 2
Pole Pattern Topology for High Torque and Low Torque Ripple

Abstract This chapter introduces two methods for effectively reducing torque ripple. The novel radial-flux surface-mounted PM (SPM) machine with quasi-cylindrical pole pattern (QCPP) is considered to be a topology structure that can effectively reduce torque ripple while achieving high average torque. The cogging torque and torque ripple of the surface-mounted machines could be greatly decreased due to the adoption of large eccentric distance and the reduction of high-order air-gap flux density harmonics. Moreover, the increase of fundamental flux component helps to improve torque generation well. Additionally, the magnet poles are semi-embedded into the rotor, and thus the sleeve in conventional SPM machines is no longer necessary, which helps to reduce the airgap size and improve the torque output further. Besides, the quasi-cylindrical magnet could offer lower rotor eddy loss and higher anti-demagnetization capability than conventional radial and eccentric pole patterns. On the other hand, Spoke-type PM machines have strong flux-focusing effect and hence high average torque; however, they generally suffer from relatively high torque ripple because of the abundant high-order magnetic harmonics of the electric machines. The hybrid irregular magnetic pole pattern composed of quasi spoke magnets and surface mounted magnets for high-speed motors is an good candidate for reducing the torque ripple while maintaining high torque for spoke-type PM machine. The quasi-Spoke-type magnets and the surface-mounted PMs with multi-direction magnetization help to improve average torque greatly. Besides, the surface-mounted magnets with parallel magnetization are conducive to reducing high-order air-gap magnetic harmonics and hence achieving smooth torque output.

2.1 Introduction

Due to the compact rotor structure, excellent dynamic response and high efficiency, surface-mounted permanent magnet (SPM) machines have been extensively used in various industrial cases [1–3]. In practical applications, especially for aircraft electric side stick, robot joints, manufacturing, measuring instruments, etc., the torque ripple may influence perceptual accuracy, motion precision, vibration and noise of the

P. Xiang et al., *Rotor Pole Pattern Topology Technology for Permanent Magnet Electric Machine*, https://doi.org/10.1007/978-981-95-2018-3_2

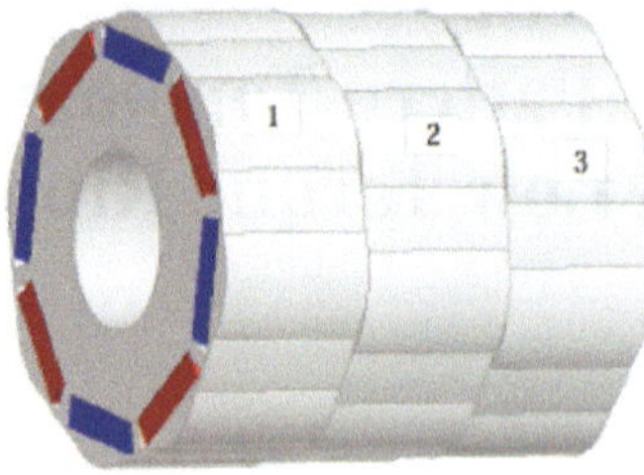

(a)

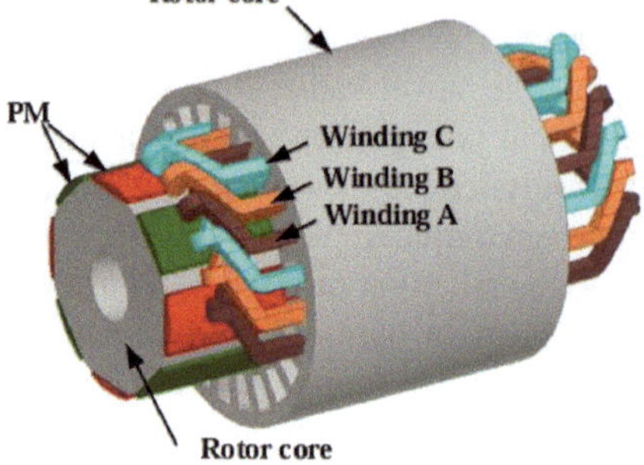

(b)

Fig. 2.1 Skewed stator slots or pole magnets. **a** Step-skewed rotor of the 12-slot-8-pole IPM machine [6]. **b** Structure of the PMSM in [7]

SPM machines greatly, and the torque output determines the loading capacity of the mechatronic systems significantly [4]. However, most existing technologies of reducing torque ripple usually result in the decrease of torque output [5].

For example, as shown in Fig. 2.1, the employment of skewing slots or pole magnets is one effective way to suppress cogging torque with optimum skew angles [6, 7]. But it generally reduces the slot fill factor and hence torque output is decreased accordingly. Furthermore, it often causes additional axial magnetic force which in turn may influence the performance and reliability of mechanical bearings [8]. Similar approaches such as changing slot width or PM arc width, shifting PM positions and notching in tooth and poles could be utilized to reduce the torque ripple through the compensation of elementary-edge cogging torque of PMs [9–11], as shown in Fig. 2.2. However, these methods may result in asymmetric airgap flux distribution and large effective airgap which in turn lead to unbalanced magnetic force and significant torque drop.

Alternatively, pole shaping methods are effective ways to reduce torque ripple, which could not generate undesired axial or unbalanced magnetic forces mentioned above. For instance, the parallel sided arc-shaped, the tapered edge shaped and the loaf-shaped magnet patterns in Fig. 2.3 are utilized to reduce the high-order harmonics of airgap flux waveform and hence suppress the torque ripple [12]. The interpolar flux leakage could also be improved. Nonetheless, those methods can only eliminate specific harmonic terms of cogging torque. Thus, one sinusoidal shaped PM pattern is studied to achieve sinusoidal airgap flux distribution [13, 14]. However, the decrease of magnetomotive force (MMF) causes significant drop of torque output. Similarly, as shown in Fig. 2.4, Zhao et al. [15] proposed one PM shape sinu-

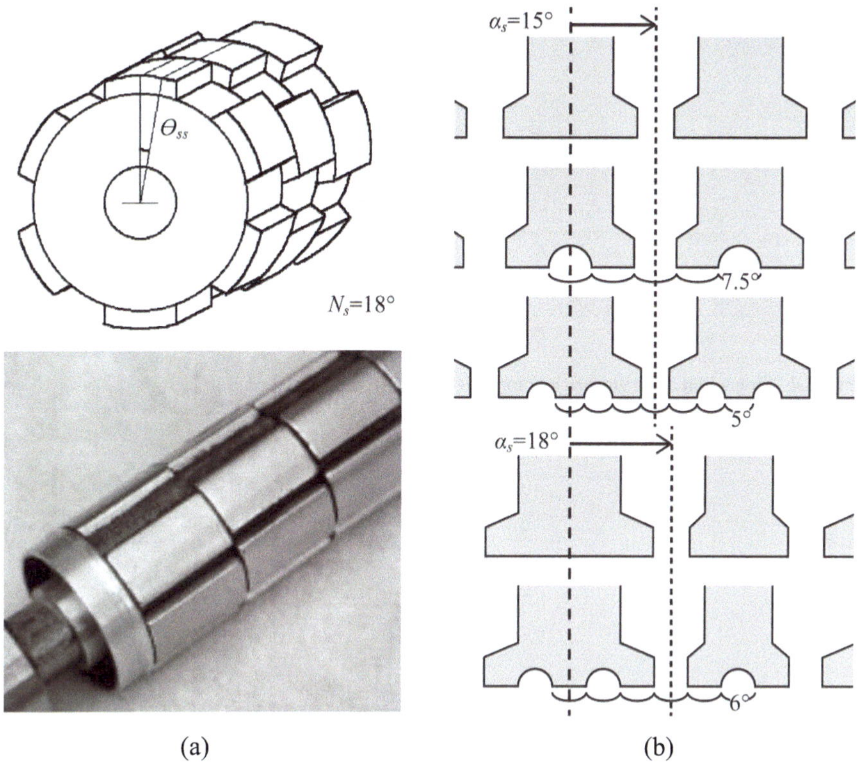

Fig. 2.2 Research models in [10, 11]. **a** Stepped rotor skewing with three modules. **b** Arrangements of actual slots and dummy slots for equal and alternate slot windings

Fig. 2.3 Loaf-shaped magnet patterns [12]

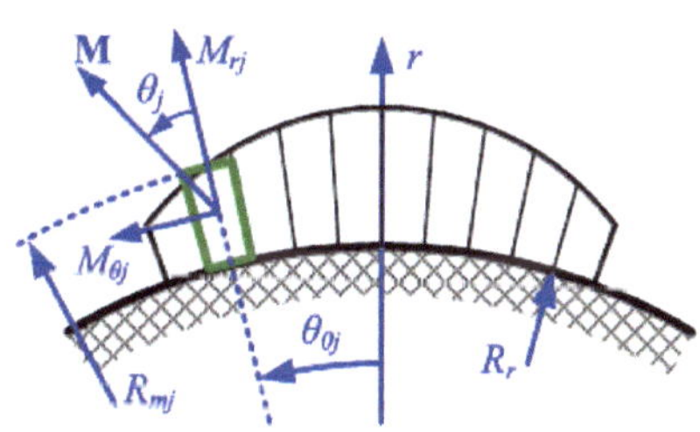

soidal and symmetrical in the axial direction to minimize the torque ripple through obtaining sinusoidal back electromotive force (EMF) in the perspective of the three-dimensional magnetic field. To reduce the total harmonic distortion and thus the torque ripple, as shown in Fig. 2.5, Upadhayay et al. [16] proposed an eccentric magnetic pole. The results shows that increasing the offset between the rotor center and the center of magnet's external surface helps to decrease the torque ripple well. This method is also employed in the development of dual-stator PM machines [17], as shown in Fig. 2.6, in which a diametrically magnetized cylindrical PM pattern

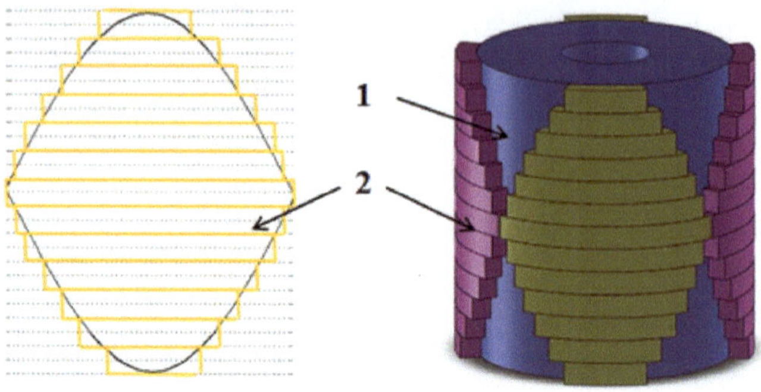

Fig. 2.4 Sinusoidal and symmetrical magnet pattern in the axial direction [15]

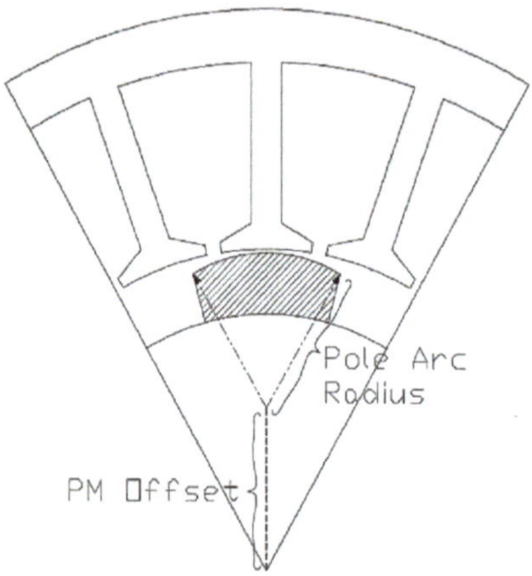

Fig. 2.5 Cross section of the 18 s/12p BLDC machine with eccentric magnets in [16]

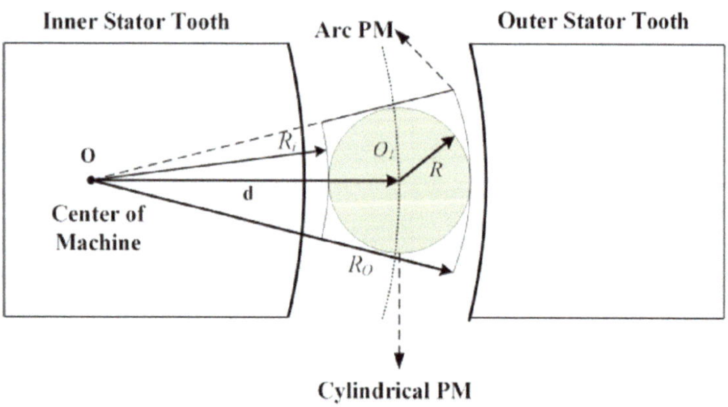

Fig. 2.6 Approximate shape of the DMCPMs and conventional machines in [17]

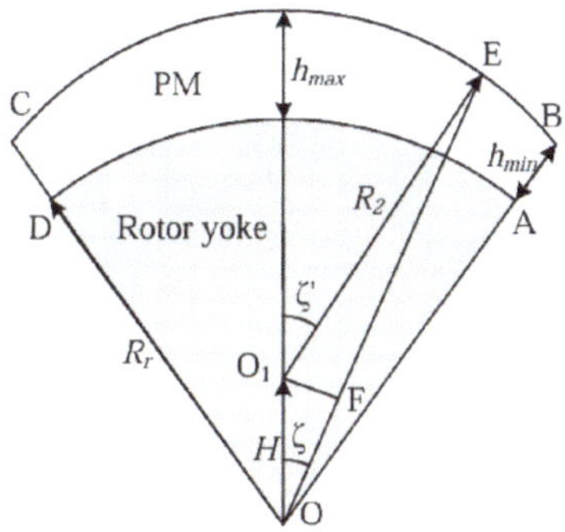

Fig. 2.7 Eccentric structure of the PM [18]

is presented. Due to the large PM offset by the cylindrical PM, the small cogging torque and torque ripple can be achieved. An analytical method is proposed in [18] to calculate the magnetic field and cogging torque of eccentric rotor shape in Fig. 2.7. However, shaping pole methods generally suffer from the PM volume decrement and the increment of equivalent air-gap length, and thus may reduce the average torque more or less. Besides, these special magnets generally need a thick rotor sleeve for reliable operation, which in turn further increases the equivalent air-gap size.

One method to overcome the problem of torque reduction is to inject flux field harmonics such as the third or fifth terms into the airgap field of dual three-phase or five-phase machines through various PM pole configuration, rotor iron flux barriers and segmented PM etc., to improve the fundamental components of airgap flux density and the torque output to some extent [19, 20]. Besides, as shown in Figs. 2.8 and 2.9, the third harmonic term can be injected into the flux field by changing PM shape in the axial direction or magnetization pattern of Halbach array to improve the

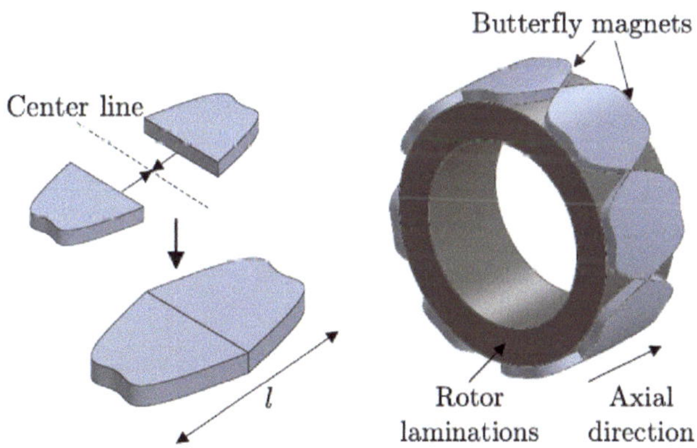

Fig. 2.8 Magnet pole pattern with butterfly shaped magnets in [21]

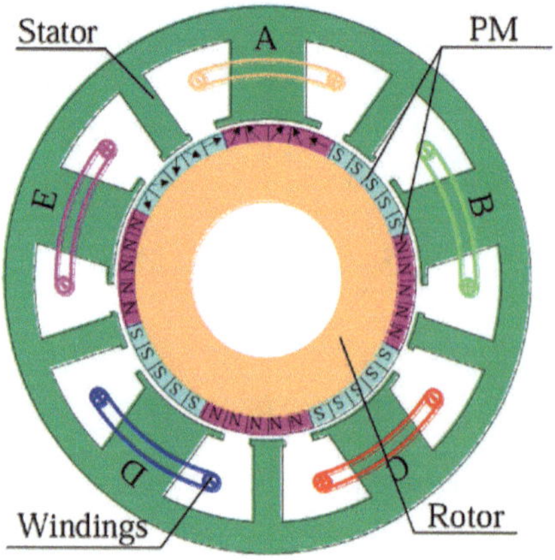

Fig. 2.9 Halbach array with magnetic harmonic injection in [22]

anti-demagnetization capability [21, 22]. However, this method ignores interpolar leakage, pole edge effect and magnetic circuit saturation etc., and the employment of high-order harmonics in the airgap flux waveform may increase the torque ripple especially for electrical machines with non-star windings [23, 24]. In fact, the harmonic injection methods could be considered as one type of pole shaping method, and thus the drawbacks of magnet shaping also exit such as the influence of rotor sleeve.

On the other hand, skewing rotor poles of Spoke-type PM machines is a simple and effective way to reduce torque ripple, but it generally leads to reduction of average torque [25]. Besides, some irregular-shaped Spoke-type pole patterns, such as trapezoidal magnets [26] and V-shaped magnets in Fig. 2.10 [27], are proposed to improve air-gap flux density distribution. However, stronger flux-focusing effect can be achieved by these approaches due to the large cross-section of these irregular

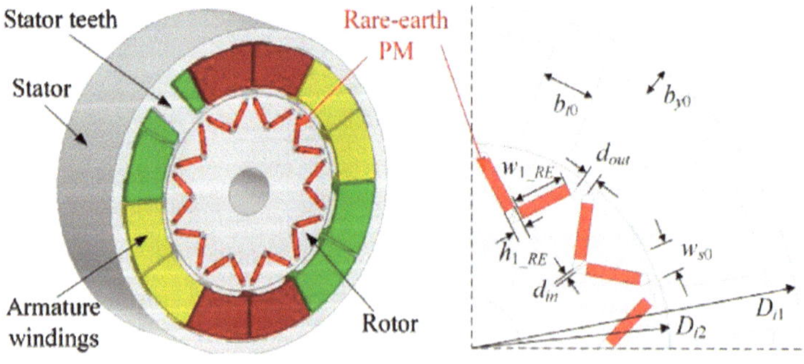

Fig. 2.10 Pole pattern with V-shaped magnets [27]

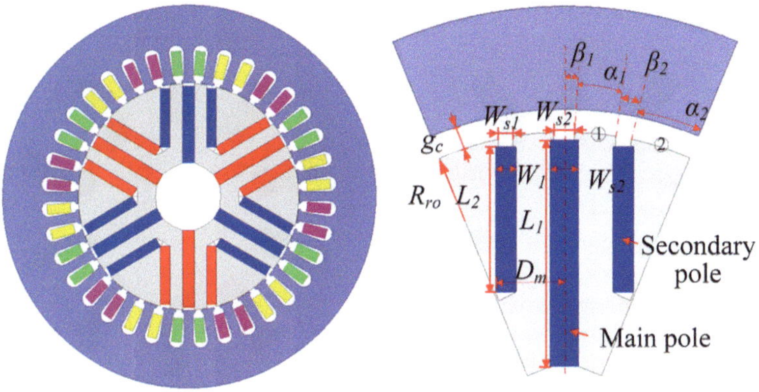

Fig. 2.11 2D Model of the two-layer spoke-type machine in [28]

Spoke-type magnets. The high-order harmonics of air-gap magnetic field are hardly reduced. In other words, the torque ripple of these electric machines is also relatively high. As shown in Fig. 2.11, Xia et al. [28] proposed a two-layer Spoke-type PM machine to improve the air-gap flux distribution and back EMF. The configuration of the muti-layer magnets helps obtain changing magnetomotive force, which contributes to achieving more sinusoidal air-gap flux density distribution compared with conventional Spoke-type machine. The muti-layer configuration is not very suitable for electric machines with large number of rotor poles. Alternatively, the torque ripple and vibration can be reduced by shaping the interpolar iron of the Spoke-type PM machines [29], as shown in Fig. 2.12. The interpolar iron with arc top in Fig. 2.13 has

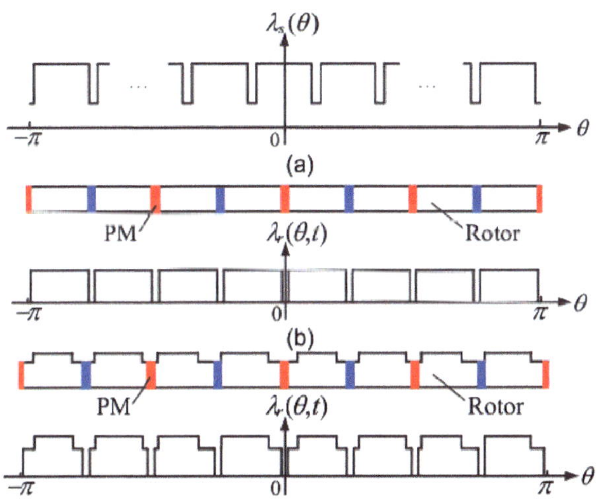

Fig. 2.12 Spoke-type PM machine with shaped interpolar iron [29]

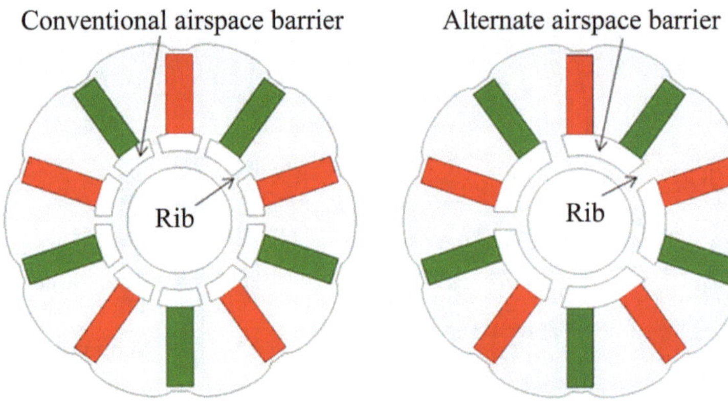

Fig. 2.13 Spoke-type IPM rotors with conventional and alternate airspace barriers [30]

the similar functions and was presented in [30]. The air-gap magnetic resistance and high-order harmonics can be both reduced contributing to obtain higher and smooth torque output. However, the protruding interpolar iron also requires large air-gap size.

Xiao et al. [31] proposed an asymmetric Spoke-type rotor topology, as shown in Fig. 2.14. An additional asymmetric flux barrier was employed in the interpolar iron for torque enhancement and torque ripple suppression. As a result, the torque ripple of 5.1% was achieved. This value of the torque ripple may be relatively high compared with surfaced-mounted counterpart. In addition, as shown in Fig. 2.15, Park et al. [32] proposed a hybrid PM machine in which three rectangular magnets with parallel magnetization are inserted into the interpolar iron between two adjacent trapezoidal Spoke-type magnets. In this case, the air-gap flux density distribution and demagnetization durability were improved to some extent. Similarly, curve-shaped magnets with parallel magnetization are mounted on the surface of the interpolar irons of the Spoke-type PM machine in [33]. The results show that higher torque

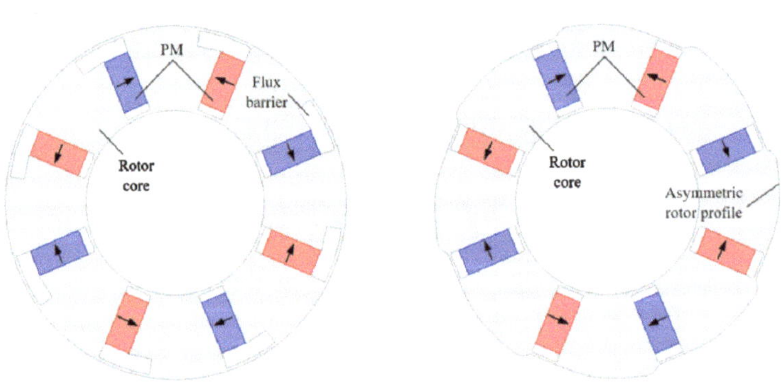

Fig. 2.14 Spoke-type IPM machine topologies with asymmetric rotor profile [31]

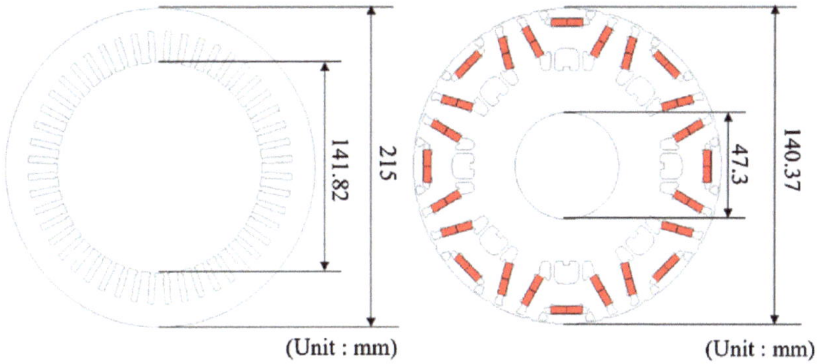

Fig. 2.15 Analysis model of the target motor in [32]

density and overload capability can be obtained compared with surface-mount and Spoke-type machines. However, the torque ripple was not suppressed well.

In general, existing methods of reducing torque ripple may cause some drawbacks. (1) The magnetomotive force may be decreased, and the equivalent airgap may be increased. As a result, the average torque output is decreased inevitably. (2) The magnet thickness is not uniform along the radial direction, which may significantly impair the anti-demagnetization capability of PM poles. (3) Additional axial or radial forces are introduced into the system due to the asymmetric magnetic field, which may cause vibration or noise unavoidably. The novel radial-flux surface-mounted PM machine with quasi-cylindrical pole pattern (QCPP) topology can depress the torque ripple effectively and improve the torque output simultaneously. The rotor design could be used in applications focusing on the performance of low torque ripple and high torque output. The utilization of large eccentric distance helps to reduce the torque ripple, and the improvement of fundamental airgap flux density aids in high torque generation. Besides, the supporting sleeve in conventional SPM machines is no longer required, resulting in a small airgap and further torque enhancement. Moreover, there is no additional axial or radial force caused by the design.

Meanwhile, under the existing spoke topology, optimization can be carried out to significantly reduce torque ripple while preserving output capability. The hybrid-irregular pole pattern is an impressive topological form for high-speed PM machines to enhance torque output capability and hence further power density. Besides, the torque ripple of the electric machine can also be suppressed greatly.

However, when designing high-performance PM machines, the design considerations are not only the fine favourable output characteristics, but also low vibration and noise. In some special equipment, the stability of the machine is also a very important aspect of inspection. For example, the high-frequency vibration of surface-mounted PM machines can affect the service life of the machine and generate a series of serious noises [38], especially under high-speed operating conditions, which is undesirable. Therefore, it is necessary and significant to investigate output characteristic and vibration response of SPM machines.

Eccentric PM is an impressive PM approach for reducing the torque ripple and vibration in machines. A novel machine with segmented PM poles and optimized Halbach array having different eccentric distance is proposed in [39], as shown in Fig. 2.16. They found that the segmented-eccentric PM poles exhibit significant potential to achieve low torque ripple, high output torque comparing with traditional Halbach array. In [40, 41], an analytical prediction modeling of air-gap flux density components in the eccentric surface-mounted PM machines is proposed, and the cogging torque and radial electromagnetic force were analyzed. Liu et al. [42] investigates the effects of eccentric PM design on sideband electromagnetic force and hence high-frequency vibroacoustic performance of integral-slot surface-mounted PM machiens. They found that the design can reduce the high-frequency vibration and noise of integral-slot PM mahines. As shown in Fig. 2.17, [43] discusses the influence of PM shape on torque characteristic in surface-mounted PM machines. The third-order harmonic PM, as shown in Fig. 2.18, was firstly used in [44] to improve the

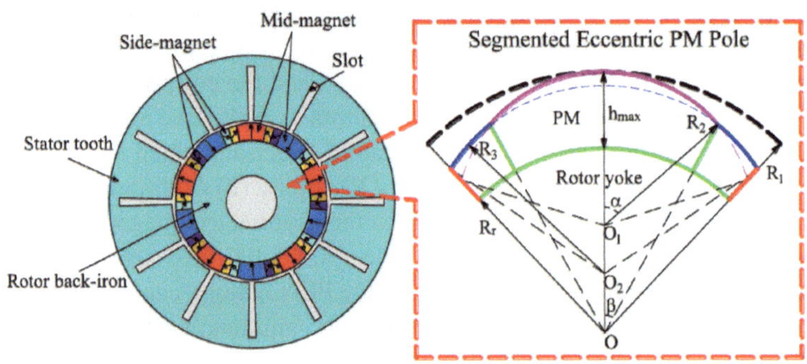

Fig. 2.16 Halbach array with eccentric rotor magnets [39]

Fig. 2.17 Magnet sine shaping and sine shaping with 3rd harmonic in [43]. **a** Sine shaping design. **b** Sine+3rd shaping design

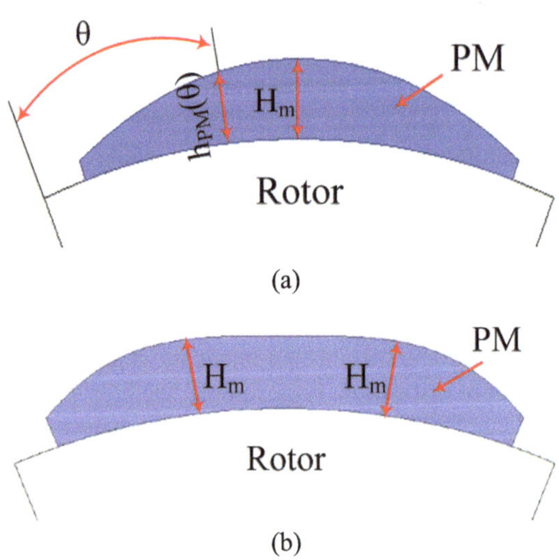

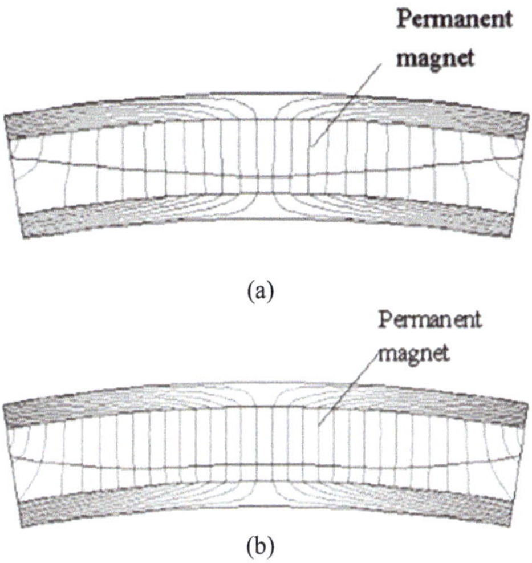

Fig. 2.18 Calculation results for sine wave designed permanent magnet in [44]. **a** Sine wave designed magnet. **b** Parallel top sine wave designed magnet

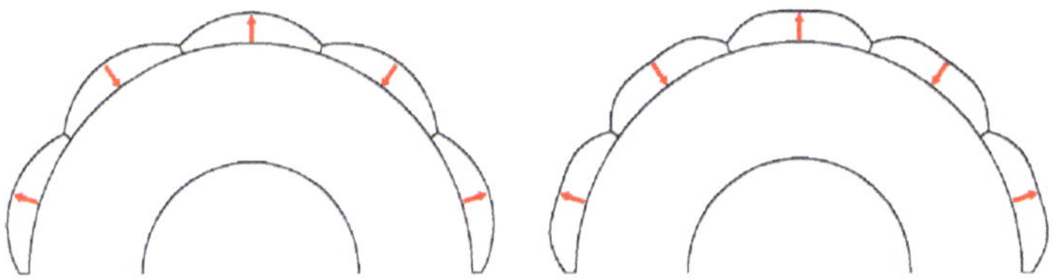

Fig. 2.19 12-slot/10-pole SPM machines with different PM shapes in [45]

efficiency of PM machines. As shown in Fig. 2.19, a PM shaping technique with optimal third-order harmonic is presented in [45] to improve the output torque. As shown in Fig. 2.20, [46] proposed a new design method for surface-mounted PM machine to reduce torque ripple. In [47], the effect of the third-order harmonic injected into the sinusoidal PM shape on the electromagnetic performance of a prototype machine is investigated, as shown in Fig. 2.21. They found that the Sin+3rd harmonic-shaped PM had a contrary effect on the torque and vibration performance of the integral-slot machien, which improved the torque capability but worsened the vibration performacne. K. Wang compared the back-EMF waveforms, cogging torque, average torque adn torque ripple for machines having rotors without shaping, Sine shaping, Sine shaping with third-order harmonic injected, as shown in Fig. 2.22. They found that the third-order harmonic PM improves torque by more than 9% in [45]. Meanwhile, the edge thickness is mentioned for the first time to ensure mechanical reliability and

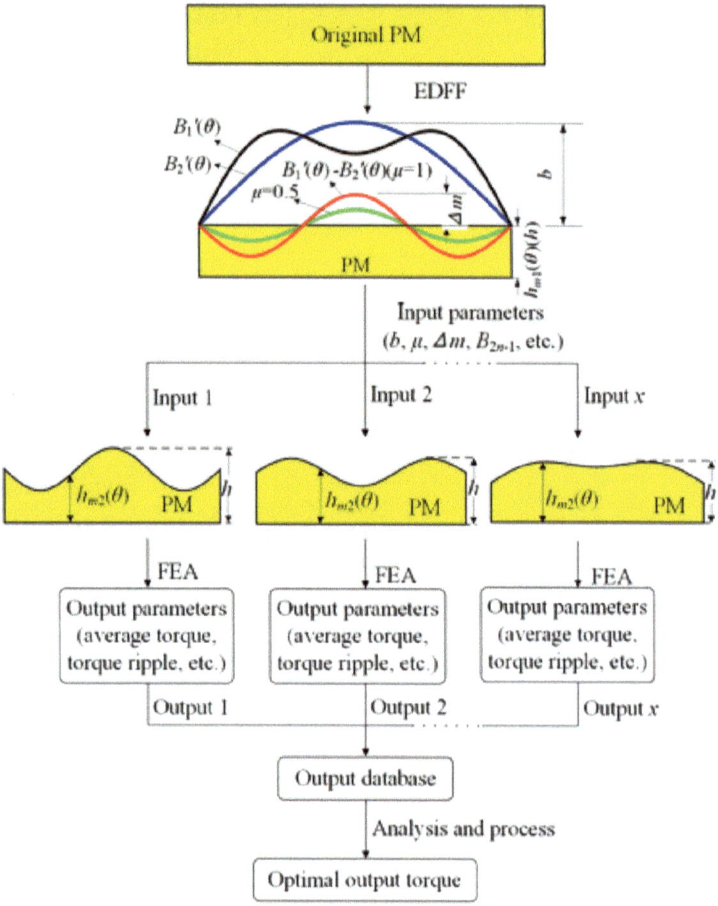

Fig. 2.20 Operating steps of EDFF in [46]

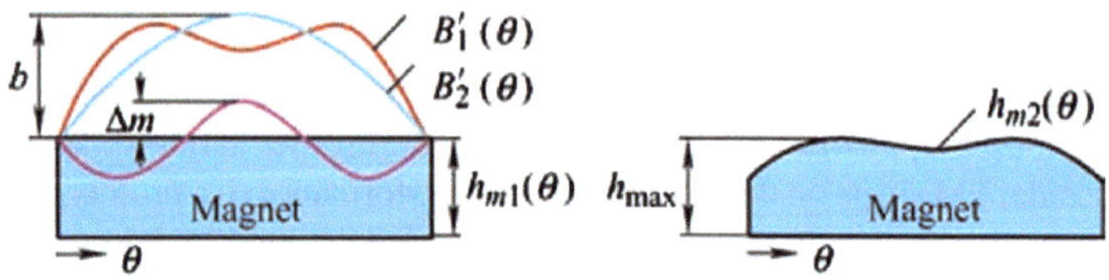

Fig. 2.21 Feedback function design applied to PM in [47]

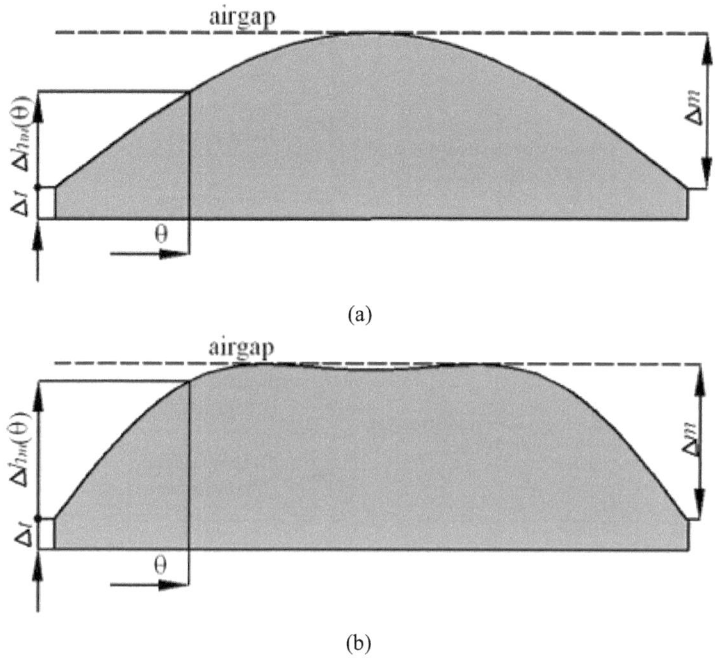

Fig. 2.22 Edge thickness compensation for Sine and Sine+3rd-shaped magnets [45]. **a** Sine shaped. **b** Sine+3rd shaped

machining feasibility. However, there is a lack of research on the influence of edge thickness on vibration and output characteristics.

The rest is organized as follows. Section 2.2.1 presents the schematic structure of the proposed machine design and its working principle. Section 2.2.2 analyzes the influence of major structure parameters on the output performance, and describes the design process of the electrical machine. In Sect. 2.2.3, the electromagnetic performance of the proposed design is analyzed and compared with other designs. In Sect. 2.2.4, one research prototype along with two test-rigs have been developed, and experiments have been conducted to validate the design concept. Section 2.3.1 presents the schematic structure of the proposed electric machine, and the characteristic of the hybrid-irregular pole pattern. Then, to obtain the optimal output performance of the electric machine, the torque profile variation with respect to the major structural parameters of the hybrid pole pattern is investigated. In Sect. 2.3.2, the electromagnetic characteristics, such as no-load air-gap flux density, back EMF and torque profile are compared with conventional designs. And then, a usage invariant third-order harmonic PM with different edge thickness is presented, and presents the derivation process of the shape function of the third-order harmonic PM for the machine. For highlight the comparison, the reserch on air-gap flux density, torque performance, radial electromagnetic force and vibration frequency response to reveal the influence of different PM edge thickness on the performance of a PM machine.

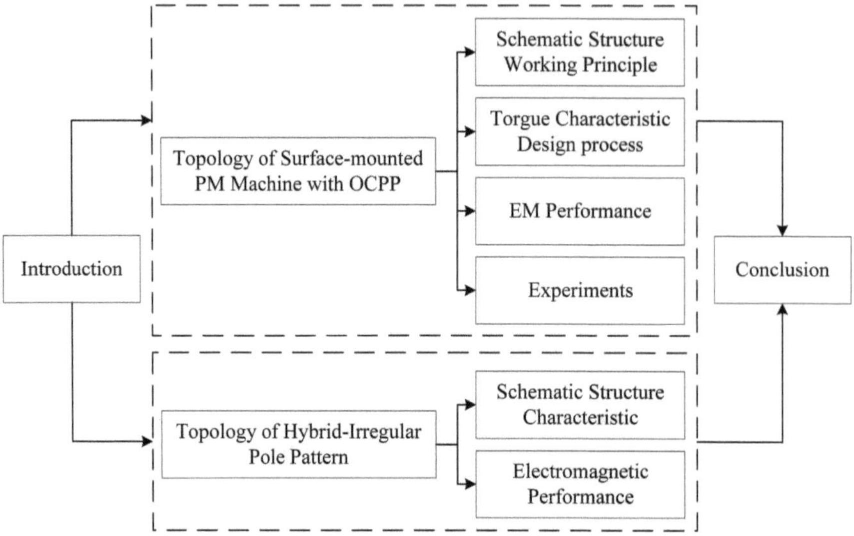

Fig. 2.23 The research content of this chapter

And suggestions were given regarding edge thickness to achieve optimal torque, torque ripple, and vibration performance. Finally, a useful conclusion is concluded. Figure 2.23 shows the scope of this chapter.

2.2 Quasi-Cylindrical Pole Pattern Topology

2.2.1 Schematic Structure and Working Principle

2.2.1.1 Schematic Structure of Quasi-Cylindrical Pole Pattern

Figure 2.24 shows the schematic structure of the proposed QCPP machine and the quasi-cylindrical pole pattern. The configuration of 18-slot/16-pole fractional slot concentrated winding is employed for suppressing torque ripple and achieving short end windings. The quasi-cylindrical pole consists of an outer arc that is eccentric to the rotor surface and a concentric inner arc. Compared with conventional eccentric pole shown in Fig. 2.25, the outer eccentric arc of the proposed magnet pole extends into the rotor yoke, forming a quasi-cylindrical shape. The proposed quasi-cylindrical pole can be divided into two parts, i.e., part 1 and part 2 as shown in Fig. 2.24b. The part 1 can be considered as conventional eccentric pole with large eccentric distance H, and hence low cogging torque can be obtained theoretically [17]. More importantly, through the part 2 in the quasi-cylindrical pole, the fundamental and third harmonic terms of the flux field can be injected into the phase electromotive force,

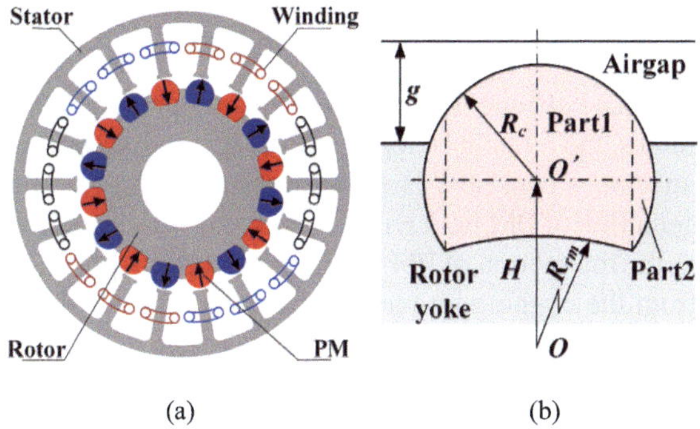

Fig. 2.24 Schematic structure of QCPP machine. **a** Cross section of 18-slot/16-pole QCPP machine. **b** Proposed quasi-cylindrical pole pattern

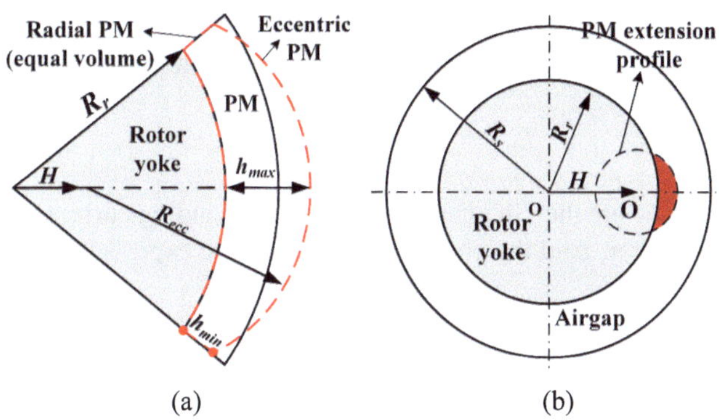

Fig. 2.25 Schematic diagram of conventional eccentric pole pattern. **a** Eccentric structure of a permanent magnet. **b** Eccentric magnet in electric machine

which helps to improve torque output greatly. In other words, the proposed pole pattern could increase average torque output and decrease torque ripple simultaneously, overcoming the drawback of conventional eccentric pole pattern well. In addition, this structure has following features:

1. The PMs of the QCPP machine have been semi-embedded into the rotor core, and thus the supporting rotor sleeve in conventional SPM machines is no longer necessary. The effective airgap size can be decreased to obtain strong flux density in airgap and hence high torque output. Besides, the eddy losses produced by sleeve is completely removed.
2. The quasi-cylindrical pole pattern can generate high magnetomotive force and hence strong anti-demagnetization capability, compared with conventional radial-magnetization and eccentric pole pattern in Fig. 2.25a.
3. The PM inner arc surface is conducive to the installation and positioning of the magnetic poles.

2.2.1.2 Working Principle

Because the difference between the direct- and quadrature-axis reactance is negligible in SPM machines, the reluctance torque can be ignored. Consequently, the torque output in SPM machines consists of following components: (1) mutual torque generated by the interaction of PM field without slots and winding currents; (2) cogging torque caused by the interaction of PM field and slot opening [34]. The torque ripple mainly comes from the cogging torque and the ripple of mutual torque. It has been proved that increasing the eccentric distance H between the center of the machine rotor and the center of outer surface of PMs could significantly reduces the cogging torque and hence torque ripple [16, 17]. As shown in Fig. 2.26, the peak cogging torque is significantly reduced by 82.3% with the increase of eccentric distance, from 8.42 mNm to 1.49 mNm. For the proposed pole pattern, the center of PM outer surface is placed at the farthest possible point from the center of the machine. To avoid an excessively large equivalent airgap, h_{min} is equal to zero and H is approximately equal to the radius of rotor, R_r. The mutual torque can be written as

$$T_{mu} = (e_{\mathrm{a}}i_{\mathrm{a}} + e_{\mathrm{b}}i_{\mathrm{b}} + e_{\mathrm{c}}i_{\mathrm{c}})/\omega_{\mathrm{r}} \tag{2.1}$$

where e_j, i_j and ω_{r} are the jth phase back EMF, the winding current and the angular velocity, respectively. And the phase back EMF can be expressed as [1]

$$\begin{aligned} e_i &= 2\omega_r R_s l_a N_{ph} \sum_{k=1,3\cdots} k_{dp} k_{sok} (B_{rck} \sin\theta - B_{rsk} \cos\theta) \\ k_{dp} &= \sin\left(\frac{k\alpha_y}{2}\right) \frac{\sin\frac{k\pi}{6p}}{q_0 \sin\frac{k\pi}{6pq_0}},\ k_{sok} = \frac{\sin\frac{kb_0}{2R_s}}{k\sin\frac{b_0}{2R_s}} \end{aligned} \tag{2.2}$$

where $q_0 = N_s/3GCD(p, N_s)$. R_s, l_a, N_{ph} and N_s are the radius of stator bore, the stack active length, the numbers of winding turn per phase and slots, respectively. K_{dp} and K_{sok} are the winding and slot opening coefficients, respectively. B_{rck} and B_{rsk} are the coefficients to be determined and closely related to the pole shape. To highlight the influence of the proposed quasi-cylindrical pole pattern shape to

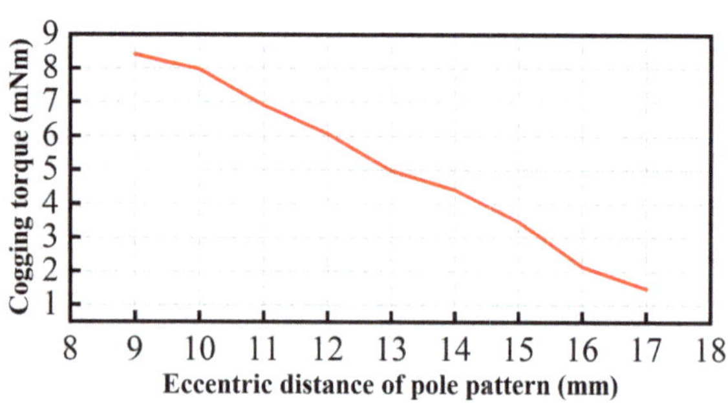

Fig. 2.26 Variation of cogging torque amplitude w.r.t eccentric distance H

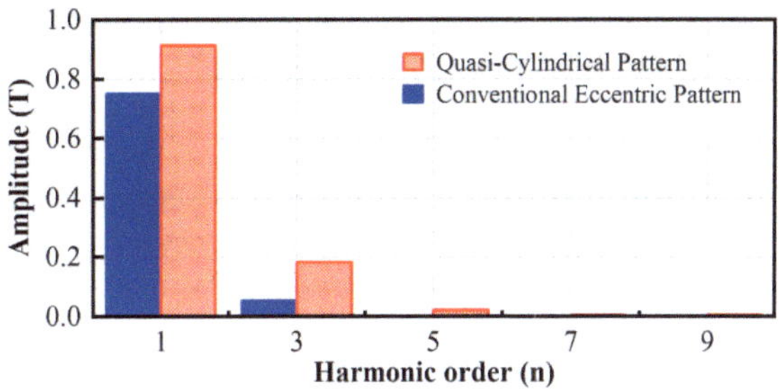

Fig. 2.27 Comparison of the amplitude of airgap flux density harmonics of quasi-cylindrical pole pattern and conventional eccentric pole pattern

the mutual torque, the effect of slot opening and winding is not considered. Then, substituting (2.2) into (2.1), the mutual torque can be given by

$$\begin{aligned} T_{mu} &= \sum_k (B_{ek} I \sin(kp\theta)\sin(p\theta) + B_{ek} I \sin(kp\theta - 2k\pi/3) \\ &\sin(p\theta - 2\pi/3) + B_{ek} I \sin(kp\theta - 4k\pi/3)\sin(p\theta - 4\pi/3)) \\ &= \frac{3}{2} B_{e1} I - \frac{3}{2} B_{e5} I \cos(6p\theta) + \frac{3}{2} B_{e7} I \cos(6p\theta) + \cdots \end{aligned} \tag{2.3}$$

where I is the phase current amplitude, and B_{ek} is the derived coefficient only related to the harmonic order k.

Equation (2.3) indicates that the torque ripple caused by the mutual torque comes from the high-order field harmonics especially the fifth- and seventh-order components of airgap flux density. The third component and its multiples do not cause torque fluctuation and the fundamental one generates main average torque. Therefore, the harmonics of phase back EMF and flux density distribution play important roles on torque output and torque ripple. Compared with conventional eccentric pole pattern with the same PM volume, the proposed quasi-cylindrical pole pattern can inject the fundamental and third components and maintain the low value of the high-order terms in airgap flux density as shown in Fig. 2.27, which helps to improve the torque output and reduce the torque ripple simultaneously.

2.2.1.3 Sizing Equation of the Quasi-Cylindrical Pole Pattern

The sizing equation for electrical machines can generally be expressed as [35]

$$P_o = \frac{1}{1 + K_a} \frac{m}{m_i} \frac{\pi}{2} K_e K_i K_p \eta B_g A \frac{f}{p} \lambda_o^2 D_o^2 L_e \tag{2.4}$$

where K_a, K_e, K_i and K_p are the ratio of electrical loading on rotor and stator, the electromotive force factor related to winding configuration, the ratio of peak and root-mean-square phase current and electrical power waveform factor, respectively.

f is the converter frequency, and p is the machine pole pairs. m and m_i is the number of machine phases and the number of phases of each stator, respectively. B_g and A are the magnetic loading,i.e., the amplitude of airgap flux density waveform, and total electrical loading, respectively. D_o is the machine outer diameter, and L_e is the effective stack length. λ_o is the ratio between airgap diameter D_g and outer diameter of machine D_o. η is the machine efficiency.

For conventional surface mounted PM machines, if its airgap flux density waveform is regarded as sinusoidal waveform, the corresponding factors in (2.4) can be simplified as follow.

$$\begin{aligned} K_a = 0,\, K_i = \sqrt{2},\, K_p = \frac{1}{2}\cos\alpha \\ A = 2m_1 N_t \frac{I_{rms}}{\pi D_g},\, m = m_i \end{aligned} \tag{2.5}$$

where N_t is the number of turns per phase. α is the phase angle difference between back electromotive force waveform and phase current waveform. Thus, the final sizing equation of conventional SPM machines is

$$P_{SPM} = \frac{\sqrt{2}}{4}\pi K_e \eta B_{gspm} A \frac{f}{p} \lambda_o^2 D_o^2 L_e \cos\alpha \tag{2.6}$$

where B_{gspm} is the peak value of airgap fundamental flux density. For the sizing equation of the proposed quasi-cylindrical pole pattern, the electrical power waveform factor K_p and the amplitude of airgap flux density waveform B_g should be rewritten according to the airgap magnetic field of the pattern. The airgap flux density of the quasi-cylindrical pole pattern is

$$\begin{aligned} B(\omega t) = \sum_k B_{rck}\cos(kp\omega t) + \sum_k B_{rsk}\sin(kp\omega t) \\ B_{rck} = \sum_j B_{rckj},\, B_{rsk} = \sum_j B_{rskj} \end{aligned} \tag{2.7}$$

where ω is the mechanical angular velocity. Therefore, the electrical power waveform factor can be expressed as

$$K_{pqc} = \frac{1}{T}\int_0^T \frac{K_e N_t B_{gqc}\frac{f}{p} D_g L_e \left(B(t)/B_{gqc}\right) \times i(t)}{E_{pk} \times I_{pk}} dt \tag{2.8}$$

where phase current $i(t) = I_{pk}\sin(p\omega t + \alpha)$. B_{gqc} is the peak value of airgap flux density. The peak value of back electromotive force is

$$E_{pk} = K_e N_t B_{gqc} \frac{f}{p} D_g L_e \tag{2.9}$$

By substituting (2.7), (2.8) and (2.9) into (2.4), the following sizing equation of the machine with quasi-cylindrical pole pattern is obtained as

$$P_{QC} = \frac{\pi\sqrt{2}}{2} K_e K_{pqc} \eta B_{gqc} A \frac{f}{p} \lambda_o^2 D_o^2 L_e \tag{2.10}$$

$$T_{QC} = \frac{\pi\sqrt{2}}{2n} K_e K_{pqc} \eta B_{gqc} A \frac{f}{p} \lambda_o^2 D_o^2 L_e \tag{2.11}$$

where the P_{QC}, T_{QC} and n are the output power, torque output and rotor speed, respectively. The sizing equation is usually used to estimate the dimensions of the electric machine roughly [35]. It can be simplified to (2.6) for actual implementations.

2.2.2 *Torque Characteristics and Design Process*

2.2.2.1 Influence of Structural Parameters on Torque Output

Figure 2.28a shows the variation of average torque and torque ripple with respect to the external radius of the QCPP magnet pole, R_c, as shown in Fig. 2.24b. It indicates that the average torque can be improved by increasing the external radius due to the growing magnetomotive force, while the torque ripple increases first and then decreases. R_c is constrained by structure dimension and the number of pole pairs.

Figure 2.28b shows the variation of average torque and torque ripple with respect to the inner arc radius R_{rm}. Large R_{rm} indicates small permanent magnet volume of the QCPP pattern which can achieve low rotor mass and hence fast dynamic performance and low manufacturing cost. However, large R_{rm} may result in poor torque output. When $R_{rm} < 19.5$mm, the torque performance does not change significantly.

Figure 2.28c shows the variation of average torque and torque ripple with respect to the rotor radius R_{yoke}. The high value of R_{yoke} can reduce torque ripple, however, resulting in poor average torque due to the growing flux leakage in rotor core. Besides, increasing R_{yoke} also causes large rotor mass and hence poor torque density.

Based on above results, the structural parameters of the quasi-cylindrical pole pattern, R_c, R_{rm} and R_{yoke}, affect the torque characteristics significantly, and they are coupled together in a certain degree. It is necessary to determine these structural parameters for the electrical machine.

2.2.2.2 Design Process

The design process is presented in Fig. 2.29. Specifically, a large value of the slot/pole number helps to increase torque output, but results in large field-changing frequency and flux density in stator teeth and hence high iron loss. It is limited by the maximum

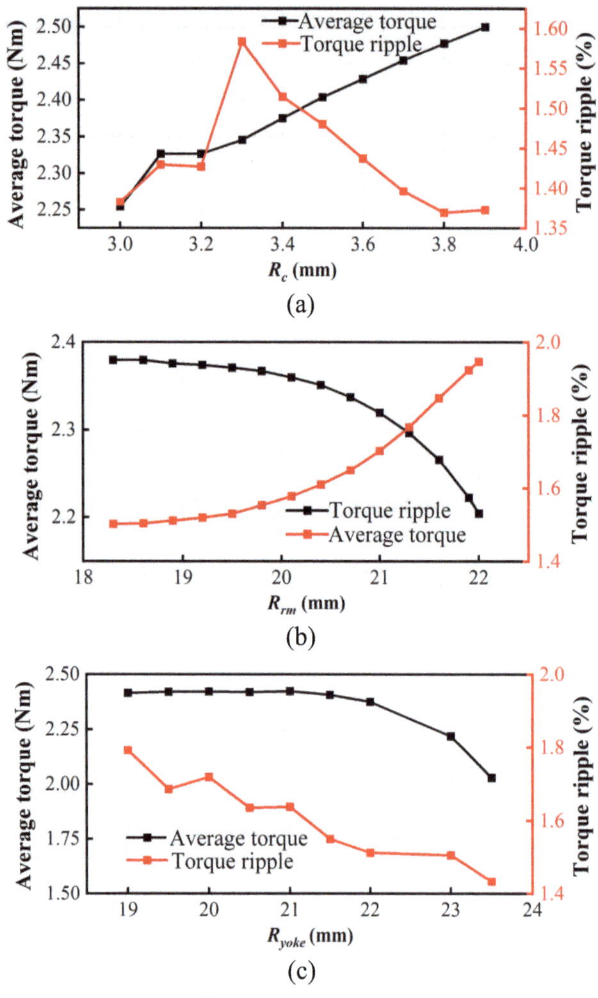

Fig. 2.28 Variation of average torque and torque ripple with respect to magnet structural parameters. **a** R_c. **b** R_{rm}. **c** R_{yoke}

stator diameter and the radius of the pole pattern R_c. According to the maximum speed and the flux density in stator teeth, the value of the slot/pole number is selected. In addition, the slot/pole numbers, the stator diameter, the active stack length of the stator and the quasi-cylindrical magnet structure parameters, R_c, R_{rm} and R_{yoke} are determined through design iteration and numerical validation. The purpose of the design is to maximize the torque output and minimize the torque ripple. The major constrains include the maximum flux density in stator teeth and rotor yoke, the maximum current density and the geometric values. The major design variables are the inner and outer radii of the stator, the number of slots and pole pairs, and the structure parameters of the quasi-cylindrical magnets, R_c, R_{rm} and R_{yoke}. The final values are listed in Table 2.1.

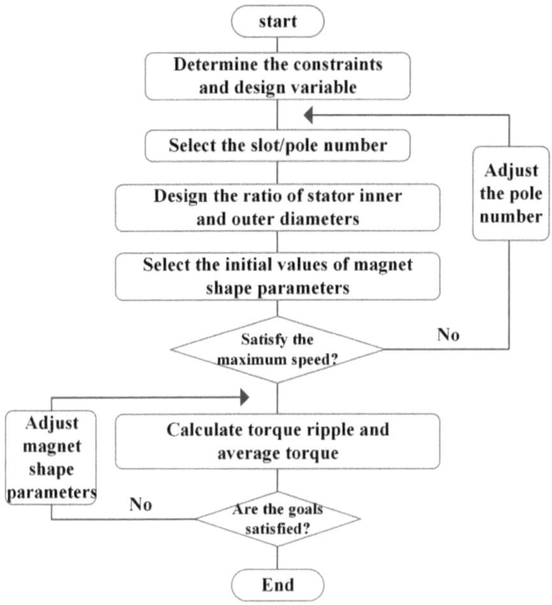

Fig. 2.29 Design process flowchart for the QCPP machine

Table 2.1 Design constraints and the final values

Symbol	Quantity	Value	Final Value
B	Magnetic flux density(max)	< 1.7T	1.5T
J	Current density	$< 6\text{A/mm}^2$	5A/mm^2
g	airgap length	< 1mm	0.8mm
D_o	Outer diameter	< 90mm	80mm
l_a	Active stack length	≤ 35mm	35mm
T_{avg}	Average torque	> 2Nm	2.4Nm
T_{cog}	Cogging torque(max)	< 0.05Nm	< 0.05Nm
T_{rip}	Torque ripple	$< 2\%$	$< 1.5\%$

2.2.3 Comparison of Electromagnetic Performance

In this section, the electromagnetic performance of the proposed quasi-cylindrical pole pattern is compared with that of the conventional radial pole pattern with optimal step slot skewing angle, the same PM volume and sleeve, and compared with that of the conventional eccentric pattern with the maximum eccentric distance equal to the radius of outer surface of rotor yoke and sleeve. The optimal slot skewing angle is determined by [10]

Table 2.2 Key design specifications of the there pole patterns

Items	Proposed pattern	eccentric pattern	Slot skewing
Number of phases	3	3	3
Outer diameter of stator (mm)	80	80	80
Inner diameter of stator (mm)	50	50	50
Stator tooth width (mm)	2.6	2.6	2.6
Airgap length (mm)	0.8	0.8	0.8
Active stack length, l_a (mm)	30	30	30
Slot skewing angle, k_{open}	–	–	10
PM grade	NdFe42	NdFe42	NdFe42
Turns per coil	45	45	45
Rated peak current (A)	6	6	6
Rated DC bus voltage (V)	48	48	48
Rated speed (r/min)	500	500	500

$$\theta_{skew} = \frac{1}{N_P}\frac{2\pi}{Q} \qquad N_P = \frac{2p}{HCF\{Q, 2P\}} \tag{2.12}$$

where $HCF\{Q, 2P\}$ represents the highest common factor between Q and $2p$. For a fair comparison, these pole patterns have the same stator dimension, rated speed and input current, as shown in Table 2.2.

2.2.3.1 Open-Circuit Flux Density in Airgap

The flux density distribution maps along with flux lines of the three pole patterns are shown in Fig. 2.30. The flux density in the stator tooth and the airgap for the quasi-cylindrical pole pattern is the highest among these patterns, because the quasi-cylindrical magnet can concentrate flux lines produced by magnets and hence improve the air gap flux density well. From the perspective of magnetic circuit, the magnetomotive force of this pattern is the highest due to the large thickness of the quasi-cylindrical magnet, which contributes to improving magnetic field and thus the torque output.

Figure 2.31a and b show the radial flux density distribution of airgap and the amplitude of corresponding harmonic components, respectively. The variation of magnetic flux density in Fig. 2.31a is consistent with the pole arrangement along the circumferential direction of the rotor. Besides, in Fig. 2.31b, the amplitude of fundamental flux component of quasi-cylindrical pattern is higher than that of the eccentric pattern. In other words, the QCPP pattern has strong flux focusing effect so that it can overcome the drawback of low MMF in conventional eccentric pole pattern, while maintaining the low high-order harmonics that helps to reduce torque ripple well.

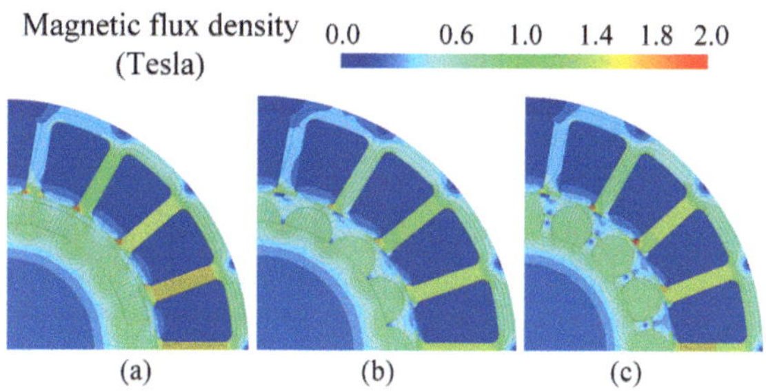

Fig. 2.30 Flux density distribution maps and flux lines of the three pole patterns. **a** Radial pole pattern. **b** Conventional eccentric pattern. **c** QCPP pattern

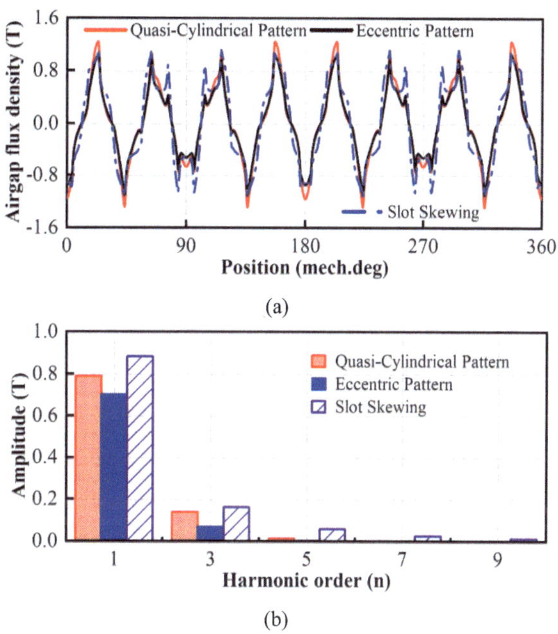

Fig. 2.31 Airgap flux density distribution of the three pole patterns. **a** Waveform of airgap flux density. **b** Amplitude of airgap flux density harmonics

The radial pole pattern with skewing slots is also investigated. The result shows that this pattern generates the highest high-order flux harmonics such as the fifth and seventh terms which do not exist in the QCPP pattern and eccentric pole pattern. These high-order harmonic terms may not only create large core loss and hence heat generation, but also result in high torque ripple.

2.2.3.2 Back Electromotive Force

The phase back electromotive force distribution and the amplitude of harmonic components are shown in Fig. 2.32. The high-order harmonics in phase back EMF of these three patterns are very low, which can generate smooth torque output with sinusoidal current input. The eccentric pole pattern has lowest fundamental harmonic

Fig. 2.32 Phase back electromotive force of the three pole patterns. **a** EMF Waveform. **b** Amplitude of EMF harmonics

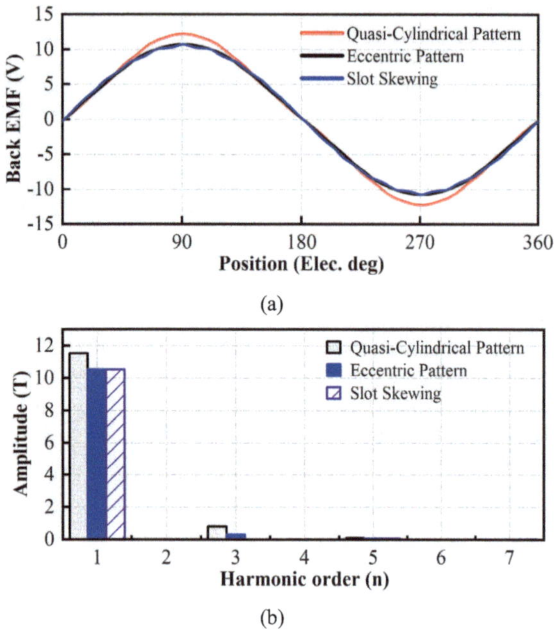

among these pole pattern because of magnet cutting and hence lower magnetomotive force. Note that the quasi-cylindrical pole pattern produces the highest fundamental amplitude than conventional eccentric and radial patterns, which helps to create large average torque output. The third harmonic component of quasi-cylindrical pole pattern is relatively higher. However, the third harmonic component has little effect on torque generation as analyzed in working principle section. Therefore, the quasi-cylindrical pole pattern can improve fundamental component and hence average torque while maintaining low torque ripple.

2.2.3.3 Comparison of Torque Performance

The torque output of these three pole patterns are investigated and compared in this section. The results are shown in Fig. 2.33. Compared with the conventional eccentric and radial pole patterns, the quasi-cylindrical pole pattern reduces the torque ripple by 17.76% and 6.02%, respectively. Meanwhile, it can increase the average torque by 9.29% and 9.04%, with respect to the eccentric and radial pole patterns, respectively. In practice, the slot fill of slot-skewing design is lower than that of other two patterns due to the difficulty of wire embedding process, which may reduce the torque output further.

The torque-angle characteristics for the three pole patterns are analyzed as shown in Fig. 2.34. The maximum average torques of the eccentric and radial pole patterns are at the point where the load angle is equal to zero. On the contrary, the maximum

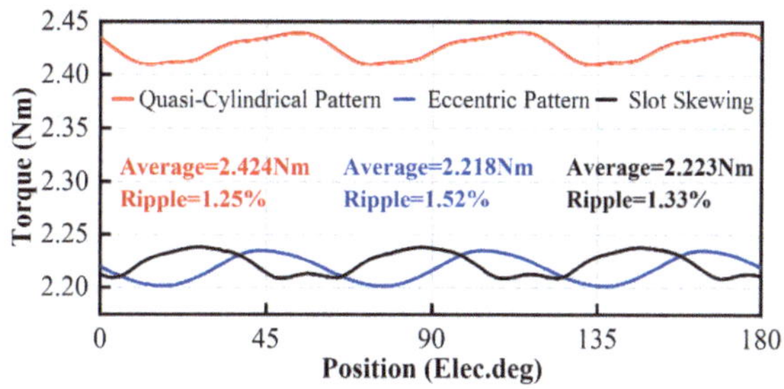

Fig. 2.33 Variation of torque output with respect to rotor position

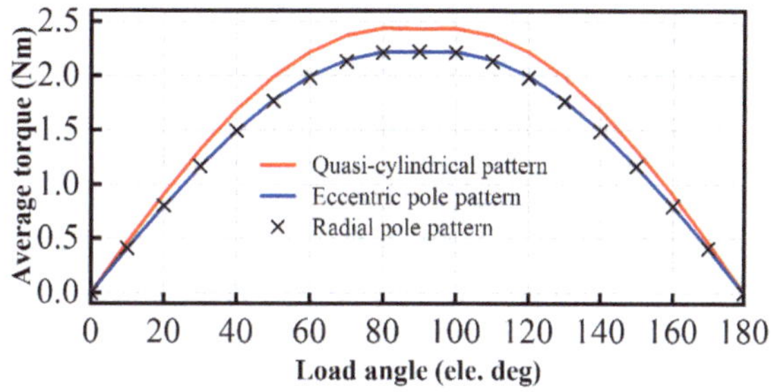

Fig. 2.34 Torque angle characteristics of the three pole patterns

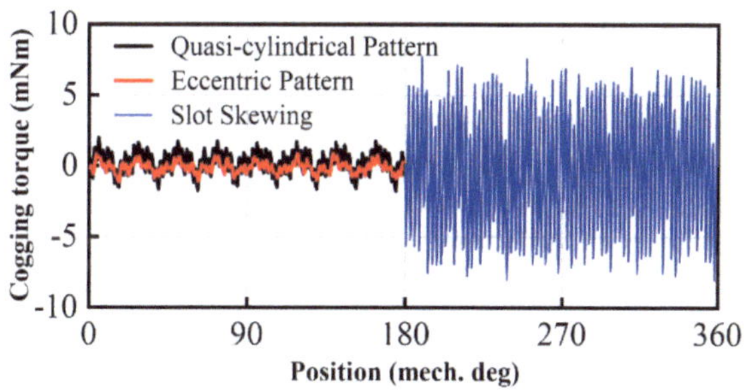

Fig. 2.35 Cogging torque of the studied three pole patterns

average torque of the quasi-cylindrical pole pattern deviates from the zero point due to the reluctance torque. In other words, the proposed pole pattern can increase the saliency ratio of the electric machine, which in turn helps to increase the constant power speed range and obtain higher average torque if an appropriate load-angle excitation is used.

Figure 2.35 shows the cogging torque of three pole patterns with NdFe35 magnets. The quasi-cylindrical pole pattern has very small amplitude of cogging torque. It is much lower than that of the radial pole pattern with skewing slots. The torque characteristics of radial pole pattern without skewing is investigated for reference. The results show that the torque ripple of the proposed cylindrical pole pattern is 22.5% lower than that of the radial pattern without skewing, and the average torque is similar.

2.2.3.4 Influence of Rotor Sleeve on Torque Output

The supporting sleeve is usually utilized to fix the magnets on the rotor of SPM machines. The most common sleeve materials are non-magnetic stainless steel or carbon fiber. The carbon fiber generally results in large airgap and hence large magnetic resistance. Thus, the thin stainless steel is often used in many electrical machines. However, the employment of stainless steel may cause magnetic short circuit that in turn reduces the torque output of the machine, because sleeve grinding process with high temperature may make rotor sleeve magnetic. Besides, the employment of stainless steel may cause eddy loss that in turn reduces the working efficiency, especially for high speed or large pole pair machines. The eddy loss also exists in the metallic sleeve and causes heat generation in rotor, which may affect anti-demagnetization capability of magnets.

To evaluate the sleeve effect quantitatively, the torque reduction and eddy loss caused by the sleeve are investigated. The 420 grade of stainless steel with low relative permeability of 40 is employed. Figure 2.36 shows the flux-line path and the eddy loss distribution of these two pole patterns. It indicates that some flux lines form close loop within sleeve, without passing through the windings. The flux field distribution results in torque drop of 32.39% for the radial pole pattern and 35.17% for the eccentric pole pattern. In addition, the sleeve eddy loss of radial pole pattern is higher than that of the eccentric pole pattern. The eddy loss may increase further for high-speed electrical machine. In contrast, the sleeve is not necessary for the quasi-cylindrical pole pattern, and thus the influence by the rotor sleeve does not exist any more. The eddy loss in the magnets of the three pole patterns is shown in Fig. 2.37. The radial pole pattern has the highest eddy loss (43mW), and the quasi-cylindrical pole pattern (29mW) and the conventional eccentric pole pattern (24mW) have similar eddy loss. The sleeve loss for the radial pole pattern and the eccentric pole pattern are 25W and 6.5W, respectively. There is no sleeve eddy loss for the quasi-cylindrical pole pattern because no sleeve is employed in the rotor. Therefore, the total loss of the proposed quasi-cylindrical pole pattern is lowest among the three magnet patterns.

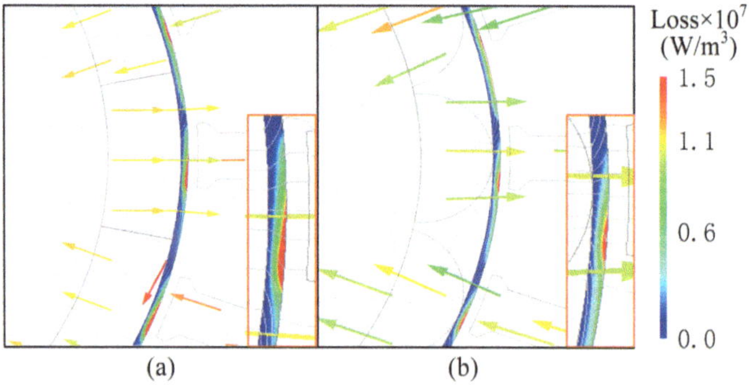

Fig. 2.36 Eddy loss of sleeve of conventional rotor designs. **a** Conventional radial pole pattern. **b** Conventional eccentric pole pattern

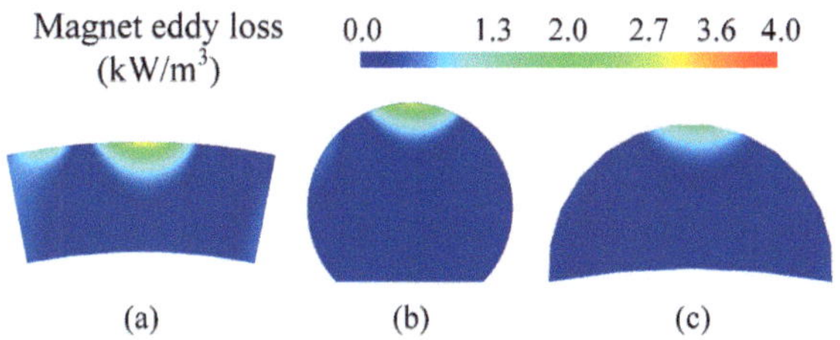

Fig. 2.37 Magnet eddy loss of the three studied pole patterns. **a** Radial pole pattern. **b** Proposed quasi-cylindrical pole pattern. **c** Eccentric pole pattern

2.2.3.5 Anti-Demagnetization Capability

The demagnetization analysis is significant for the design of pole pattern, especially for the eccentric magnet pattern. The large eccentric distance can decrease the cogging force, but the thickness of the left and right regions of magnet in Fig. 2.38 is also reduced, which results in inevitable demagnetization effect and the compromise of torque output. The demagnetization of magnet for the three pole patterns is studied under overload condition ($i_d = -3\sqrt{2}I_{rms}$). Fig. 2.39 shows the demagnetization characteristic of magnets in this study. The knee point of the demagnetization curve is about at (–0.5T, –1200 kA/m). The demagnetization of PM can be evaluated by the ratio coefficient $k_d = (1 - B_r^{\vdash}/B_r^{\dashv}) \times 100\%$, where $B_r^{\dashv}$ and $B_r^{\vdash}$ are the remanence of PM before and after irreversible demagnetization, respectively.

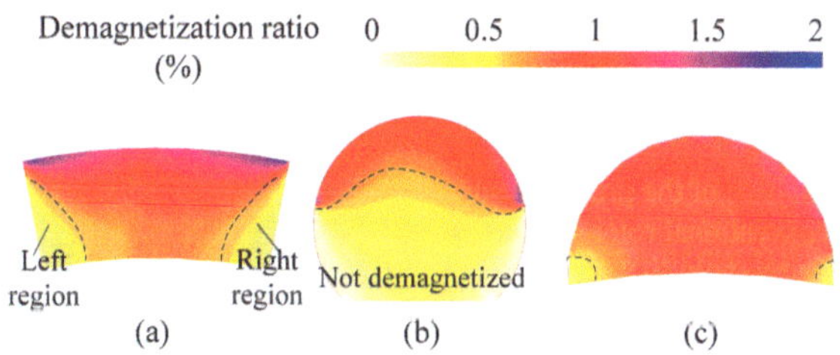

Fig. 2.38 Demagnetization analyses results. **a** Conventional radial pole pattern. **b** Proposed quasi-cylindrical pole pattern. **c** eccentric pole pattern

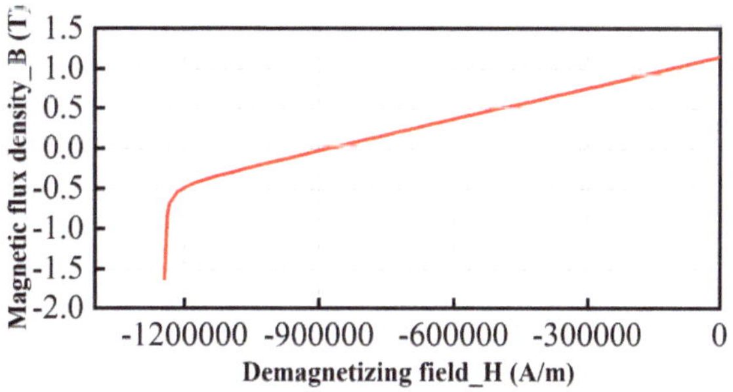

Fig. 2.39 Magnetic flux density versus demagnetizing field of the magnet

The distribution of the demagnetization ratio of the three pole patterns is presented in Fig. 3.39. The high-degree demagnetization mainly occurs at the top edge of PM close to the airgap, especially for the radial pole pattern. On the contrary, the low-degree demagnetization exits at the left and right regions of PM because the flux density of these regions is mainly determined by small flux leakage rather than high armature-reaction field by i_d. The demagnetization area of the quasi-cylindrical pole pattern is the smallest among the three pole patterns. Therefore, the quasi-cylindrical pole pattern has good anti-demagnetization capability suitable for field-weakening control for large constant power range.

2.2.3.6 Rotor Stress Analysis

To ensure the proposed rotor design can work reliably, the rotor stress analysis at high speeds is conducted. Figure 2.40 shows the von-Mises equivalent stress distribution map of the conventional radial pole pattern (without sleeve) and the proposed rotor pattern (without sleeve) under the speed of 50000rpm. The maximum magnet stress of the radial pole pattern is much higher than that of the quasi-cylindrical pole pattern. Their maximum stresses are lower than the tensile yield strength 80MPa of the magnets. However, the maximum stress of the structural adhesive in the conventional rotor structure is higher than its tensile yield strength 20MPa. On the contrary, the maximum stress of the structural adhesive in the proposed rotor structure is much lower than its tensile yield strength 20MPa. In other words, the traditional rotor structure has the falling risk of permanent magnets if there is no sleeve protection, while the proposed rotor structure can withstand the centrifugal force at high speed by the structural adhesive and the semi-embedded structure without a sleeve.

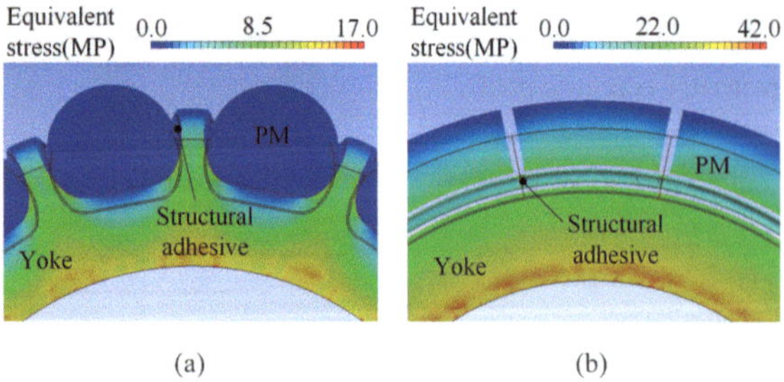

Fig. 2.40 Rotor stress analysis of the proposed and conventional rotor designs. **a** Quasi-cylindrical pole pattern. **b** Conventional radial pole pattern

2.2.4 *Research Prototype and Experiments*

2.2.4.1 Development of Research Prototype

The 18-slot/16-pole machine with quasi-cylindrical PM pattern has been developed for experiments, as shown in Fig. 2.41. The 18-slot stator and the 16-pole rotor with quasi-cylindrical PMs are shown in Fig. 2.41a and b, respectively. The parallel-magnetized quasi-cylindrical PMs are semi-embedded into the rotor core, so that the magnets are firmly fixed and the supporting sleeve can be removed completely. To measure the flux density distribution of the airgap, the front cover of the machine is made hollow so that the Hall probe can be inserted, as shown in Fig. 2.41c. The cost of the quasi-cylindrical magnets is similar to the cost of commonly used tile magnets.

2.2.4.2 Experiments and Discussion

The radial airgap flux density is measured by using the Hall probe with thickness of 0.25mm, as shown in Fig. 2.42. The experimental results in Fig. 2.43 agree with the numerical simulation well. The flux-density curve is characterized by flat-topped and spires waves caused by large slot openings and the third harmonic term, respectively. As anticipated before, the third harmonic term has less effect on torque output, and the large slot opening could result in cogging torque.

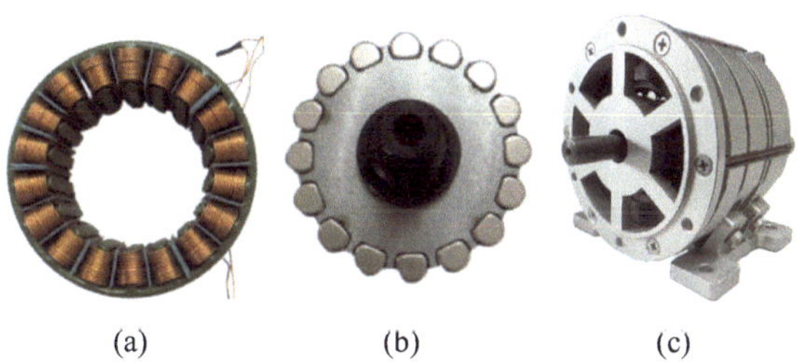

Fig. 2.41 Developed prototype. **a** Stator. **b** Rotor. **c** Assembly

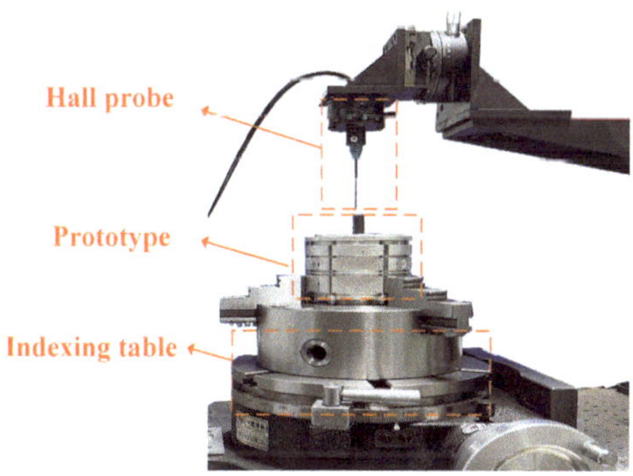

Fig. 2.42 Constructed test platform for airgap flux density distribution

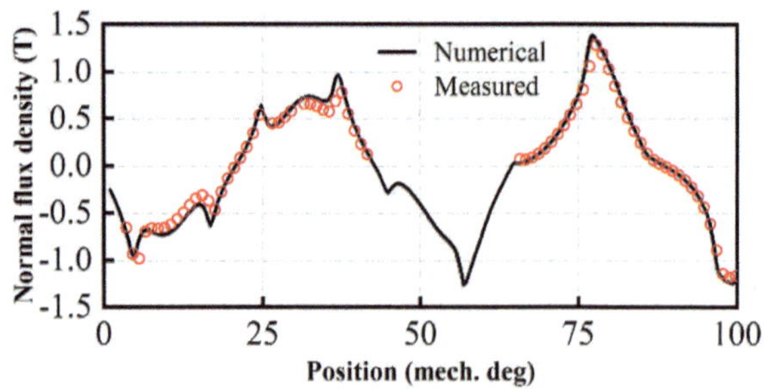

Fig. 2.43 Comparison of numerical and measured airgap normal flux density

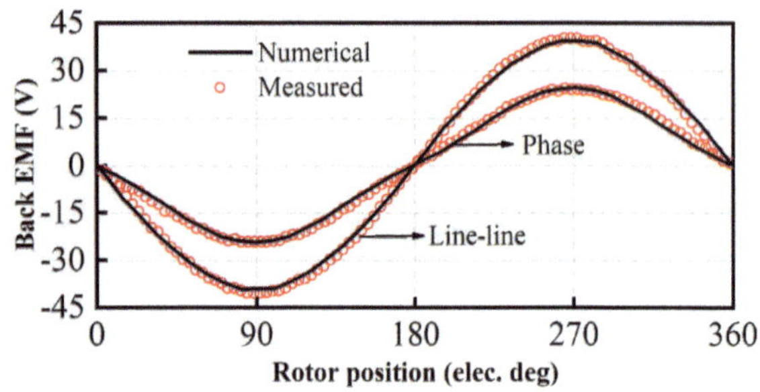

(a)

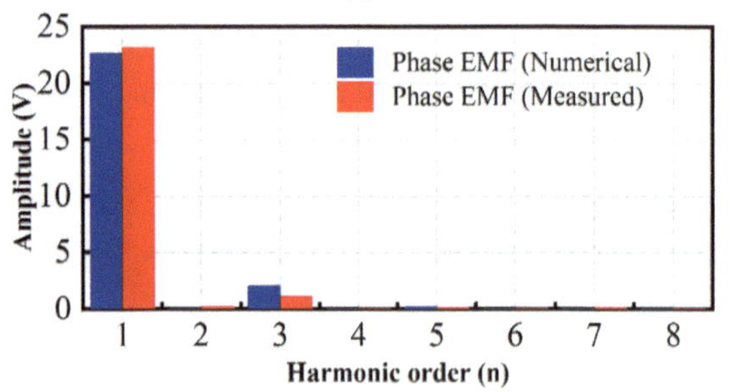

(b)

Fig. 2.44 Numerical and measured back electromotive force. **a** Waveform of phase and line-to-line back EMF. **b** Amplitude of back EMF harmonics

The output performance such as average torque and torque ripple without input current harmonics disturbance introduced by inverter is directly determined by the phase EMFs. To measure the back EMF, the prototype is driven to rotate at the rated speed by the prime mover, and the winding terminal voltage is measured by an oscilloscope. The experimental result is shown in Fig. 2.44, and it is consistent with the numerical result. The phase back EMF hardly contains high-order harmonic terms, which contributes to generate smooth torque output.

Figure 2.45 shows the constructed test platform on torque output with respect to rotor position. The output shaft of the research prototype is connected with a dynamometer through a disc coupling. The prototype generates the rotary motion and pulls the dynamometer to rotate synchronously. Meanwhile, the dynamometer can produce loading torque on the prototype. The corresponding torque output, rotor

Fig. 2.45 Constructed experimental platform for torque output performance

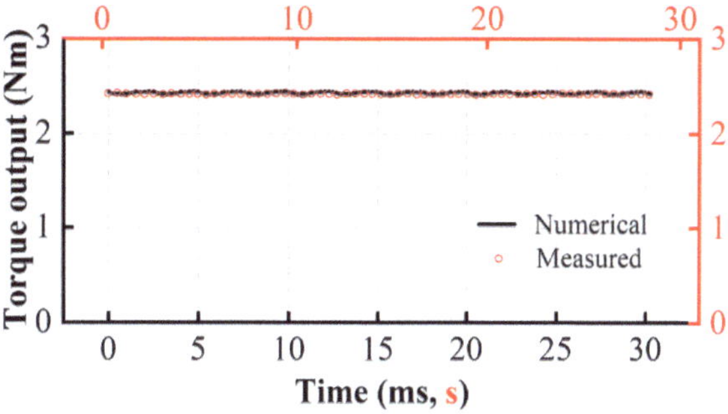

Fig. 2.46 Torque output variation of the Quasi-cylindrical pattern w.r.t. time

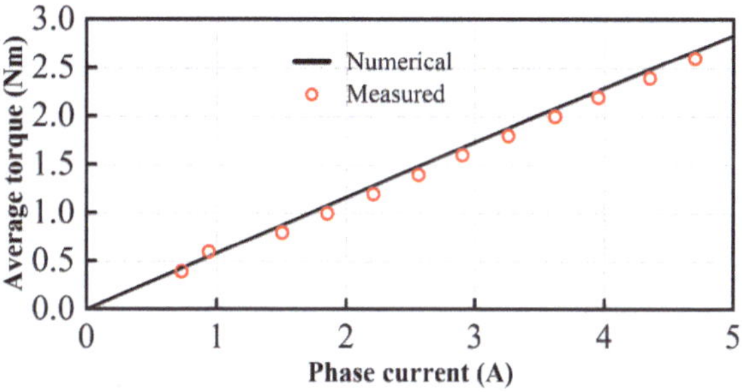

Fig. 2.47 Average torque variation of the research prototype w.r.t. current

speed and position are measured in real time by sensors integrated in the dynamometer. The measured torque output at rated speed and current input are presented in Fig. 2.46. It indicates that the experimental results of average torque agree with the numerical simulation well, and the electric machine with quasi-cylindrical pole pattern can generate very smooth torque. In Fig. 2.47, the average torque variation with respect to phase current is presented. Good agreement between numerical and measured results is achieved, which in turn validates the proposed machine design, theoretical and numerical analyses.

2.3 Hybrid-Irregular Pole Pattern Topology

2.3.1 Schematic Structure and Characteristic

2.3.1.1 Schematic Structure of the Hybrid-Irregular Pole Pattern

Figure 2.48 shows the schematic structure of the high-speed machine with hybrid-irregular pole pattern. The configuration of 18slots/4poles is employed for low operation frequency and hence low core loss. The hybrid-irregular pole pattern consists of quasi-T-shaped magnets with tangentially magnetization, curve-shaped magnets with parallel magnetization and rotor irons, as shown in Fig. 2.48. Besides a rotor sleeve is employed to protect magnets and irons under high-speed operation of the electric machine. The interaction between the current of the three-phase stator winding and the air-gap magnetic field produced by the hybrid-irregular pole pattern produces the torque and speed outputs of the electric machine.

2.3.1.2 Characteristic of Hybrid-Irregular Pole Pattern

The strong flux-focusing effect and sinusoidal air-gap flux density can be obtained by the hybrid-irregular pole pattern. The former helps to improve torque and power density, and the latter is conducive to reducing torque ripple. Specifically, magnetic flux of each pole is generated by five magnet segments, which contributing higher fundamental component of air-gap magnetic field and hence higher torque output. In addition, the air-gap flux density distribution can be improved by modifying θ_p of curve-shaped magnets. It is worth noting that the span of the top magnet of the quasi-T-shaped magnet will also change with the change of θ_p to ensure that even if θ_p is very small, the hybrid-irregular pole pattern can obtain strong flux-focusing effect. In this case, the high torque output and low torque ripple can be achieved by the pole pattern.

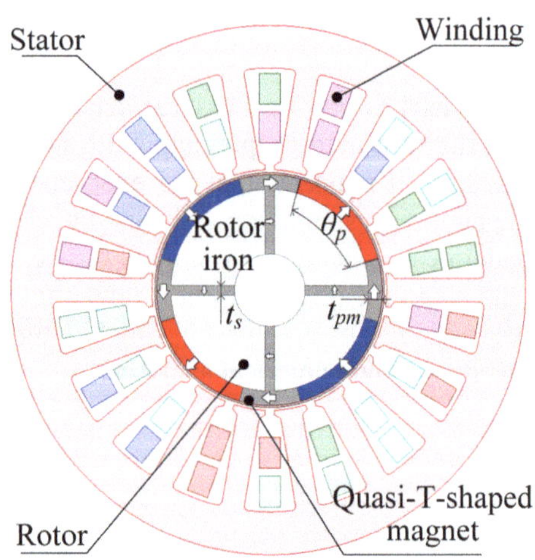

Fig. 2.48 Structure of the hybrid-irregular pole pattern

2.3.1.3 Influence of Structural Parameters

Figure 2.49a shows the variations of average torque and torque ripple with respect to magnet thickness of surface-mounted magnets, t_{pm}. With the increase of t_{pm}, the magnetomotive force of the hybrid-irregular pole pattern increases resulting higher average torque. However, it also leads to higher torque ripple. Figure 2.49b indicates that the torque ripple of the electric machine is minimum when t_s equals to 2mm. In addition, the larger t_s is, the larger flux-focusing effect is, which is beneficial to increasing the average torque. Besides, when the value of t_s is much small, the torque ripple is relatively high. And when t_s exceeds 2mm, the torque ripple will also increase to a certain extent. These indicates that the quasi-T-shaped magnets not only helps to increase the flux-focusing effect and hence high average torque,

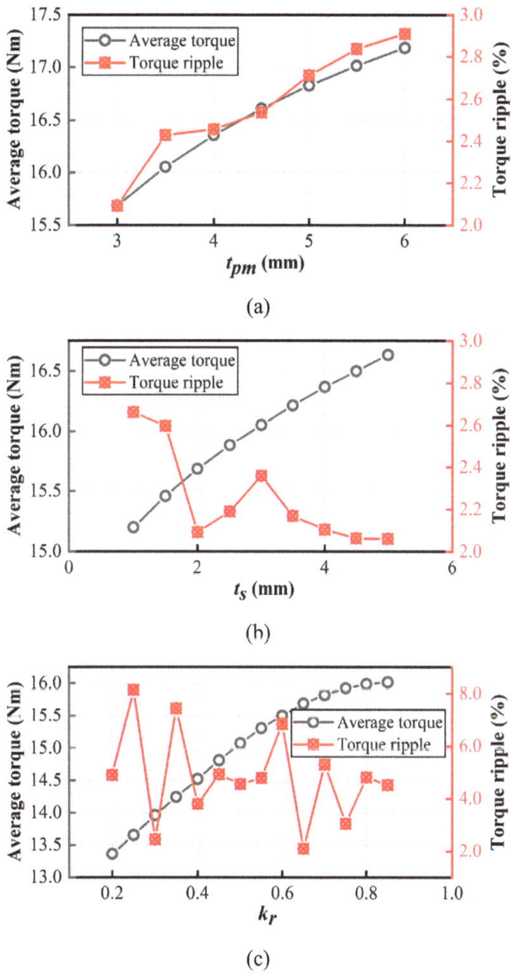

Fig. 2.49 Variations of average torque and torque ripple with respect to structural parameters. **a** t_{pm}. **b** t_s. **c** k_r

Table 2.3 Major design specification of electric machines

Items	Values	Units
pole patterns	Spoke-type/Hybrid-irregular	/
Number of phases	3	/
Outer diameter of stator	110	mm
Inner diameter of stator	50	mm
Stator tooth width	4.8	mm
Airgap length	1	mm
Active stack length	140	mm
Magnet grade	N42UH	/
Stator and rotor core grade	JFE_10JNEX900	/
Rated peak current	101	A
Rated speed	15000	r/min

but also improves the air-gap magnetic field distribution to a certain extent. The variations of average torque and torque ripple with respect to pole-arc coefficient k_r of curve-shaped magnets are shown in Fig. 2.49c. It is found that the torque ripple reaches the minimum when $k_r = 0.65$. And the average torque is also relatively high. In addition, too large and too small k_r all result in large torque ripple, because the large k_r may cause flat top waveform of air-gap flux density and small k_r may lead to peak waveform of air-gap flux density. In general, a larger k_r will help improve average torque of the electric machine. To obtain high average torque and low torque ripple under small magnet volume, the t_{pm}, t_s and k_r are set to 3mm, 2mm and 0.65, respectively. The other major design specification of the electric machine with the hybrid-irregular pole pattern is listed in Table 2.3.

2.3.2 *Comparison of Electromagnetic Performance*

To confirm the advantages of the proposed pole pattern, the electromagnetic performance of the electric machine with hybrid-irregular pole pattern, such as air-gap magnetic field, back EMF and torque profile are compared with conventional Spoke-type design. For a fair comparison, the two pole patterns have the same magnet usage, current input, and stator structure.

2.3.2.1 Open-Circuit Air-Gap Magnetic Field

Figure 2.50 shows the comparison of magnetic field between the conventional Spoke-type design and the hybrid-irregular pole pattern. Figure 2.50a and b indicate that the proposed pole pattern has stronger flux-focusing effect. In addition, Fig. 2.50c shows

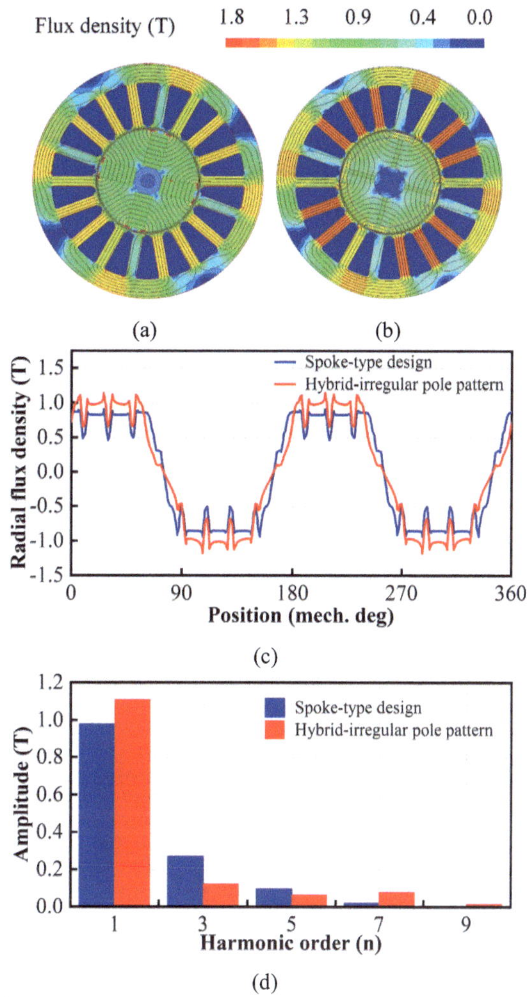

Fig. 2.50 Comparison of magnetic field. **a** Spoke-type design. **b** Hybrid-irregular pole pattern. **c** Air-gap flux density. **d** Harmonics of air-gap magnetic field

that the Spoke-type design has lower amplitude of air-gap flux density waveform, and richer high-order magnetic harmonics, such as the 3rd and 5th terms in Fig. 2.50d, than the hybrid-irregular pole pattern. In other words, the air gap magnetic field distribution of the hybrid-irregular pole pattern is closer to a sinusoidal waveform, which helps to obtain smaller torque ripple and iron loss under high-speed case than the conventional Spoke-type machine. In addition, the hybrid-irregular pole pattern has higher fundamental term of air-gap flux density distribution than that of the Spoke-type design.

2.3.2.2 Back Electromotive Force

Figure 2.51 shows the comparison of back electromotive force between the conventional Spoke-type design and the hybrid-irregular pole pattern. It is clearly found that the hybrid-irregular pole pattern has the highest amplitude of back EMF which is conducive to obtaining higher average torque. Besides, the harmonic terms of the back EMF of the two pole patterns are compared in Fig. 2.52. The results indicate that the values of the fundamental terms of the conventional and the novel design are 148.14V and 164.25V, respectively. In other words, the 10.87% increase in the fundamental term of back EMF is achieved. In addition, due to the addition of the surface-mounted magnets with muti-direction magnetization, there are less high-order harmonics, such as 3rd and 5th, can be found in the hybrid-irregular pole pattern, which contributing to torque ripple suppression. The reduction of the less high-order magnetic harmonics also helps to decrease stator core loss especially in the high-speed electric machine.

2.3.2.3 Torque Profile and Power Density

The waveforms of torque output of the Spoke-type and hybrid-irregular pole patterns are shown in Fig. 2.53. The average torque of the electric machine with hybrid-irregular pole pattern is increased by 15.54% form 13.58Nm to 15.69Nm, which

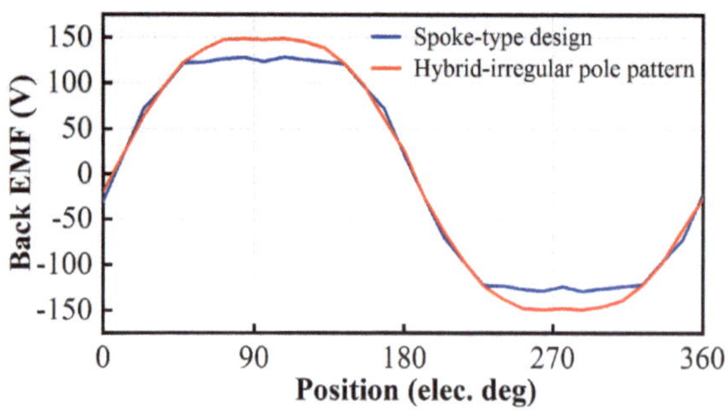

Fig. 2.51 Comparison of back EMF of the two pole patterns

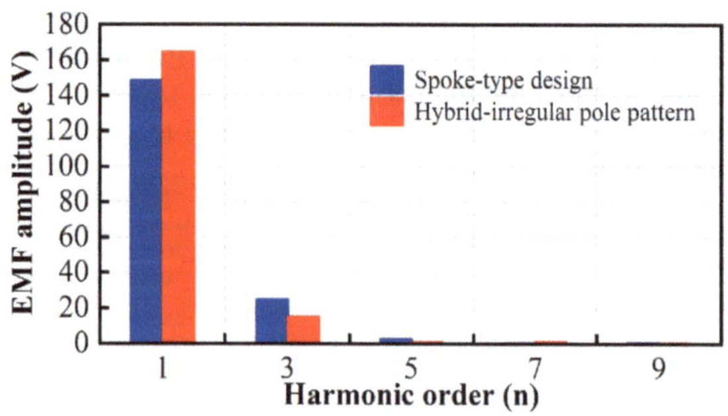

Fig. 2.52 EMF amplitude of the two studied pole patterns

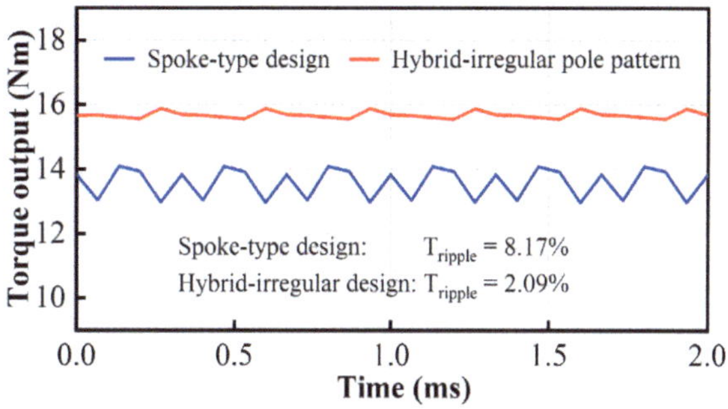

Fig. 2.53 Comparison of torque output of the two pole patterns

verifies the stronger flux-focusing effect of the proposed design compared with Spoke-type design. In addition, the torque ripples of the design and conventional pole patterns are 2.09% and 8.17%, respectively. In other words, the torque ripple is reduced by 74.42%. The stator and rotor masses are about 7.027kg and 1.989kg, respectively. Thus, the total effective mass is 9.016kg, and the torque density of the Spoke-type and hybrid-irregular pole patterns are 1.51Nm/kg and 1.74Nm/kg, respectively. The output power of the two compared pole patterns calculated by $T_{avg} \times n/9550$ where T_{avg} is average torque. In this case, the output power of the Spoke-type and designed pole patterns are 21.33kW and 24.64kW, respectively. The power density of the electric machine with hybrid-irregular pole pattern is 2.73kW/kg that is 15.19% higher than 2.37kW/kg of the Spoke-type machine. The detailed output characteristics of these two pole patterns are listed in Table 2.4.

Figure 2.54 shows the variations of average torque and torque ripple of the Spoke-type PM machine with respect to current angle. It is clearly found that the maximum average torque is 14.07 Nm (current angle = 20 deg). However, the torque ripple is also increased from 8.17% to 8.37%. In this case, the average torque of the electric machine with hybrid-irregular pole pattern (15.69 Nm) is also higher than that of the conventional Spoke-type pole pattern.

Table 2.4 Output performance comparison

Items	fSpoke-type	Hybrid-irregular
Average torque (Nm)	13.58	15.69
torque ripple (%)	8.17	2.09
Stator mass (kg)	7.027	7.027
Rotor mass (kg)	1.989	1.989
Total effective mass (kg)	9.016	9.016
Ouput power (kW)	21.33	24.64
Torque density (Nm/kg)	1.51	1.74
Power density (kW/kg)	2.37	2.73

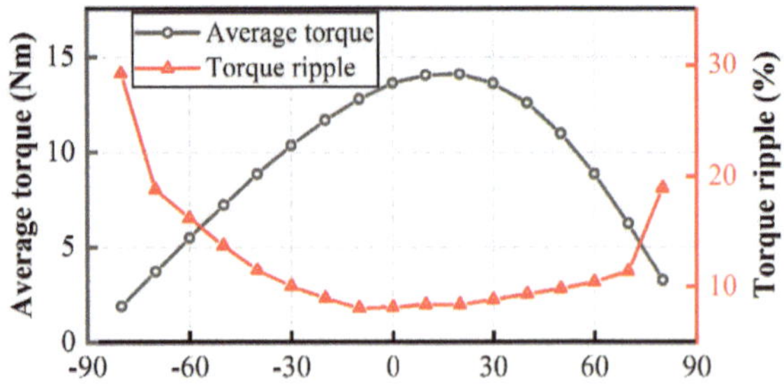

Fig. 2.54 Variations of average torque and torque ripple of the Spoke-type PM machine with respect to current angle

2.3.2.4 Losses and Efficiency

Figure 2.55shows the total loss distribution of stator cores of the two pole patterns. For the both pole patterns, the largest total loss is located at the stator tooth tips because the magnetic field changes here are strong. In addition, although the magnetic flux density in the stator core of the hybrid-irregular pole pattern, as shown in Fig. 2.50, is higher than the Spoke-type design, the total loss of the stator core of the hybrid-irregular pole pattern is lower than that of the Spoke-type design. This is because the air-gap magnetic field of the hybrid-irregular pole pattern has less high-order magnetic harmonic content. Figure 2.56 shows the rotor total loss distribution of the two compared pole patterns. It is found that the main total loss of the hybrid-irregular pole pattern occurs on the surface of the permanent magnets close to the air-gap. The total losses of the rotor irons of the two compared pole patterns are both very small.

The PM eddy-current loss, core loss and copper loss of the two pole patterns are calculated and compared in Fig. 2.57. As anticipated in Fig. 2.56, the PM eddy loss of the hybrid-irregular pole pattern is higher than that of the conventional Spoke-type design. This is because there are surface-mounted magnets close to the air-gap in the hybrid-irregular pole pattern. On the contrary, the core loss of the proposed rotor design is lower than that of the conventional one, although stronger magnetic field

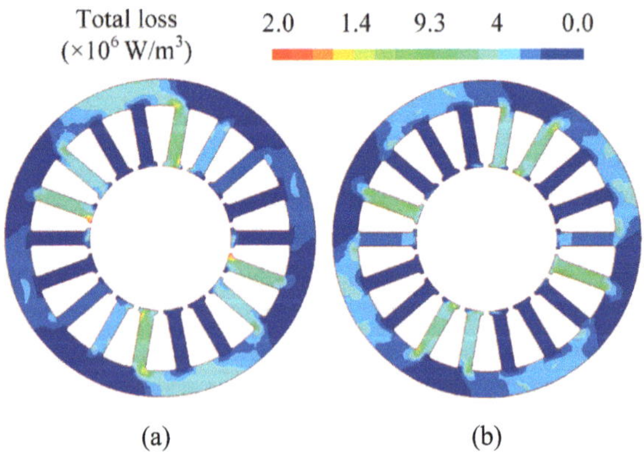

Fig. 2.55 Comparison of total loss of stator core. **a** Spoke-type design. **b** Hybrid-irregular pole pattern

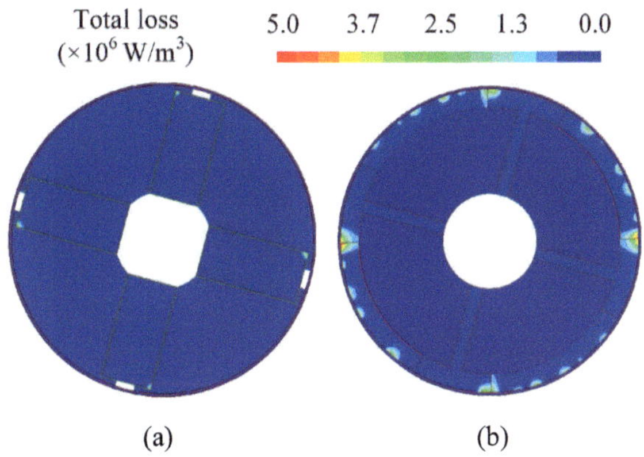

Fig. 2.56 Comparison of total loss of rotor. **a** Spoke-type design. **b** Hybrid-irregular pole pattern

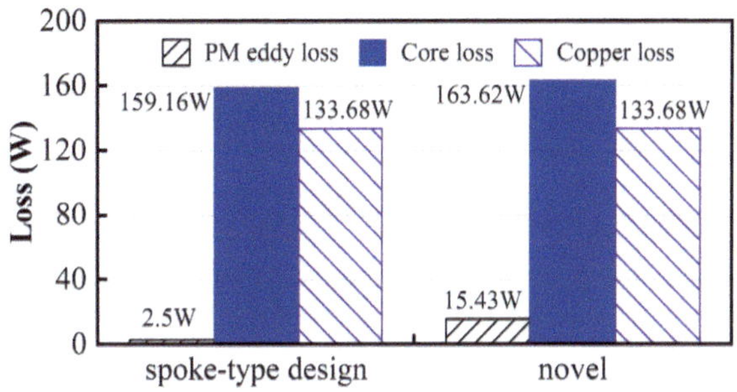

Fig. 2.57 Comparison of losses of the two pole patterns

can be found in the hybrid-irregular pole pattern. The efficiency of the both electric machines are calculated by

$$\eta = \left(\frac{P_{\text{out}}}{P_{\text{out}} + P_{\text{core}} + P_{\text{cu}} + P_{\text{m}} + P_{\text{stray}}} \right) \times 100\% \tag{2.13}$$

where P_{out}, P_{core}, P_{cu}, P_{m} and P_{stary} are output power, core loss, copper loss, magnet eddy-current loss and stray loss. P_{stary} is assumed to be 1% of the output power [36]. In this case, the efficiency of the Spoke-type and hybrid-irregular pole pattern designs are 89.78% and 97.78%, respectively. In other words, compared with the conventional Spoke-type design, the working efficiency of electric machines is improved greatly with the hybrid-irregular pole pattern.

2.3.2.5 Unbalanced Magnetic Force

To ensure the steady operation of the high-speed electric machine, the unbalanced magnetic force of the two compared pole patterns is calculated. The force terms of unbalanced magnetic force, F_{x} and F_{y}, acting on a rotor are [37]

Fig. 2.58 Unbalanced magnetic forces of the two pole patterns

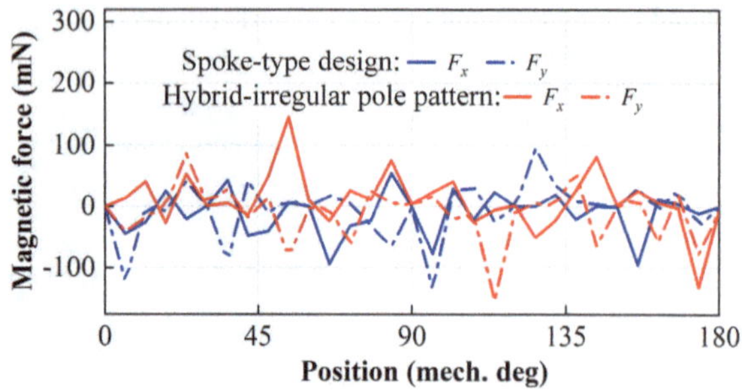

$$F_x = \frac{rl_a}{2\mu_0} \int_0^{2\pi} [(B_\alpha^2 - B_r^2)\cos\alpha + 2B_r B_\alpha \sin\alpha] d\alpha \quad (2.14)$$

$$F_y = \frac{rl_a}{2\mu_0} \int_0^{2\pi} [(B_\alpha^2 - B_r^2)\sin\alpha - 2B_r B_\alpha \cos\alpha] d\alpha \quad (2.15)$$

where B_α and B_r are the tangential and radial components of air-gap flux density distribution. μ_0 and l_a are the vacuum permeability and active stack length. The unbalanced magnetic forces of the two pole patterns are shown in Fig. 2.58. The compared pole patterns have similar and low unbalanced magnetic force which contributes to the steady operation of the high-speed electric machine.

2.4 Harmonic-Injected Pole Pattern Topology

abstractHarmonics can be injected into the permanent magnet (PM) shape to improve the output performance of PM machines, and the PM edge thickness is introduced to ease manufacture. This section investigates the effects of third-order harmonic PM with different edge thickness on magnetic field, electromagnetic force and vibration performance of integral-slot surface-mounted PM machine. Firstly, the shape functions of third-order harmonic PM with different edge thickness are derived and validated in this section. Moreover, the air-gap flux density and torque characterisitc are analyzed to obtain the potential impact of PM edge thickness on output performance. Then, the radial electromagentic force and its 2D-fast Fourier transformation (FFT) are analyzed and compared to obtain the order and frequency of the highest amplitude radial electromagnetic force, and the effect of different PM edge thickness on radial electromagnetic force. Finally, the vibration synthesis model is used to predict the vibration frequency response of machines with different PM edge thickness. The results shows the PM edge thickness can deteriorate the low-frequency vibration of machines, and it can supress the high-frequency vibration when the selection of PM edge thickness is approoriate.

2.4.1 Models and Specifications

A 24-slots/4-poles PM machine with third-order harmoncis PMs is designed for investigation, as shown in Fig. 2.59. In order to reveal the performance of PM machine with different PM edge thicknesses, the shape of the PM as shown in the Fig. 2.60 was obtained. The main parameters of the PM machine and details of the third-order PM are listed in Table 2.5.

Fig. 2.59 Model and flux density distribution of the 24-slots/4-poles PM machine

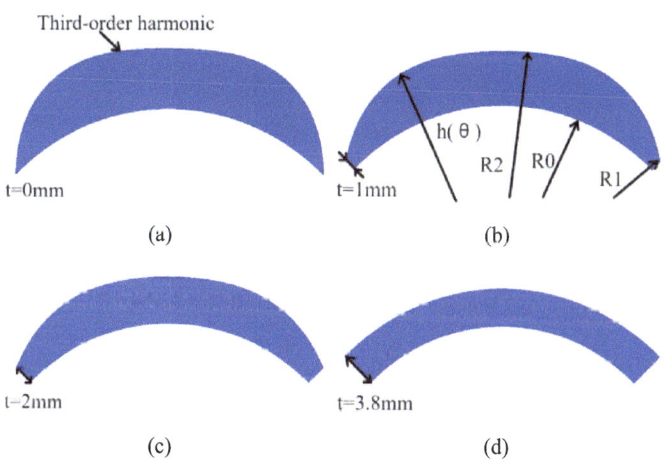

Fig. 2.60 The models of third-order harmonic PM with. different edge thickness. **a** 0 mm. **b** 1 mm. **c** 2 mm. **d** 3.8 mm

Table 2.5 Main parameters of PM machine

Symbol	Quantity	Value
p	Number of pole-pair numbers	2
Q	Number of stator slots	24
C	Current density	6.2A
V	DC link voltage	270V
S	Rated speed	15000rpm
P	Rated output power	21kW
L	Air gap	1.5mm
R_1	Radius of third-order harmonic injection	R_0+t
R_2	Radius of the PM outer surface	23.5mm
R_0	Radius of the rotor inner surface	Undetermined
t	edge thickness	Undetermined

In this case, the aim is to obtain a third-order harmonic PM with the same usage as the conventional surface-mounted. The air-gap flux density distribution of third-order PM is given by

$$B_g(\theta) = a(\sin(p\theta) + b\sin(3p\theta)) \tag{2.16}$$

Assuming that the relative permeability of PM is the same as the airgap and the permeability of the iron is infinite, ignoring the influence of stator slot, the air-gap flux density can also be expressed as

$$B_g(\theta) = \frac{h_m(\theta)B_r}{L} \tag{2.17}$$

where B_r is the magnet remanence of PM, $h_m(\theta)$ is the thickness of PM.

In third-order harmonic PM, the top of the PM is defined by the third-order harmonic function which can be expressed as

$$h(\theta) = R_1 + aL[\sin(p\theta) + b\sin(3p\theta)] \tag{2.18}$$

where $h(\theta)$ is the variation in the thickness of the top of the PM. R_1 and R_2 are parameters to be determined, which are limited by the usage of PM and the edge thickness. The usage of PM is limited as

$$V = \frac{1}{2}\int_0^{\frac{\pi}{p}} h(\theta)d\theta + \frac{\pi(R_1^2 - R_0^2)}{4} = V_0 \tag{2.19}$$

where the value of V_0 is 128, which is determined by the usage of conventional surface-mounted PM. The output average torque can be obtained as

$$T_m = \frac{3e_a i_a}{2\omega_r} = \frac{3}{2} i_2 R_s l_a N_p K_{dp1} \frac{B_r(R_2 - R_1)}{L} \tag{2.20}$$

where R_s, l_a, N_p, ω_r are the airgap radius, stator axial length, the number of turns per phase, and the rotate speed.

The torque and thickness limitations of permanent magnets can be expressed as:

$$\begin{cases} \frac{dh_m(\theta)}{d\theta} = 0 \\ \frac{dT_m}{db} = 0 \end{cases} \tag{2.21}$$

As shown in Fig. 2.60d, the maximum edge thickness (the thickness of conventional PM) is 3.8 mm, so analysis will be conducted at 1mm intervals. Analyze and compare the machine performance when the edge thickness is 0 mm, 1 mm, 2 mm, 3 mm, and 3.8 mm (conventional). According to the above limiting conditions, the third harmonic function based on different edge thickness can be expressed as

$$\begin{cases} h_1(\theta) = 18.40 + 5.889[\sin(2\theta) + \frac{1}{6}\sin(6\theta)], t = 0\,mm \\ h_2(\theta) = 19.73 + 4.353[\sin(2\theta) + \frac{1}{6}\sin(6\theta)], t = 1\,mm \\ h_3(\theta) = 21.08 + 2.794[\sin(2\theta) + \frac{1}{6}\sin(6\theta)], t = 2\,mm \\ h_4(\theta) = 22.43 + 1.235[\sin(2\theta) + \frac{1}{6}\sin(6\theta)], t = 3\,mm \end{cases} \tag{2.22}$$

2.4.2 Comparison of Electromagnetic Performance Under Different Edge Thickness

Before the electromagnetic properties analysis, several regulations need be made to eliminate the influence of other factors.

1. The material of all PMs are N42UH.
2. The usage of PMs is same to ensure the fairness in comparison.
3. To eliminate the influence of magnetic saturation, stator is made of Hiperco 50 material to ensure that magnetic saturation will not occur under rated operating condition.

From the derivation, the overall third-order harmonic PM can be obtained. However, this shape of PM is difficult to achieve in processing, so the edge thickness is required. In order to reveal the influence of PM edge thickness on the electromagnetic and output characteristics of the machine, various properties of the machine will be analyzed based on numerical.

2.4.2.1 The Comparison and Analysis of Air-Gap Flux Density and Torque

Figure 2.61a shows the air-gap flux density distributions for third-order harmonic PM with different edge thickness. It is clearly found that the sinusoidal characterisitc of conventional surface-mounted PM is worst, approximating square waves while ignoring the influence of stator slots. On the other hand, as the edge thickness decreases, the sinusoidal characteristics of flux density become better. This is because the injection ratio of the third harmonic in the original permanent magnet increases, more fully reflecting the role of the third harmonic. As displayed in Fig. 2.61b, the conventional surface-mounted PM has the highest fundamental one whilst edge thickness of 1mm has the lowest total harmonic distortion (THD). The THD can be calculated as

$$\mathrm{THD} = \sqrt{\sum_{i=2}^{n} \frac{H_i^2}{H_1}} \times 100\% \tag{2.23}$$

where H_i is the ith order component. The fundamental component increases with the increase of edge thickness while the THD first decreases and then increases. This phenomenon is because as the edge thickness increases, the content of high-order

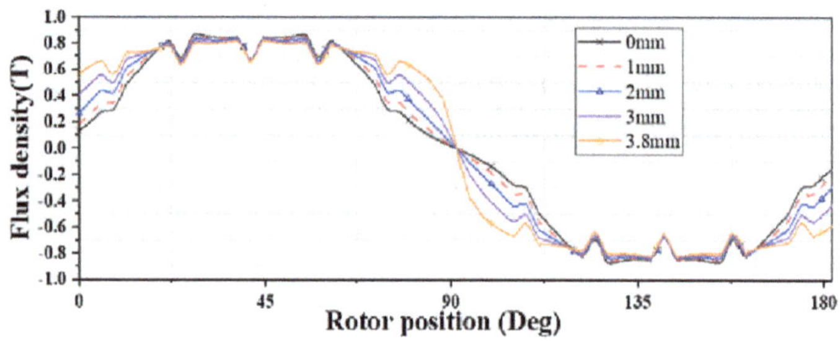

(a)

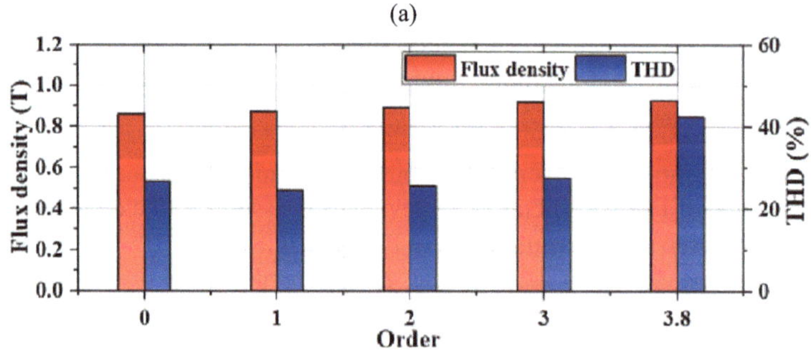

(b)

Fig. 2.61 Comparison of air-gap flux density with different PM edge thickness. **a** Flux density distribution. **b** Fundamental and total harmonic distortion

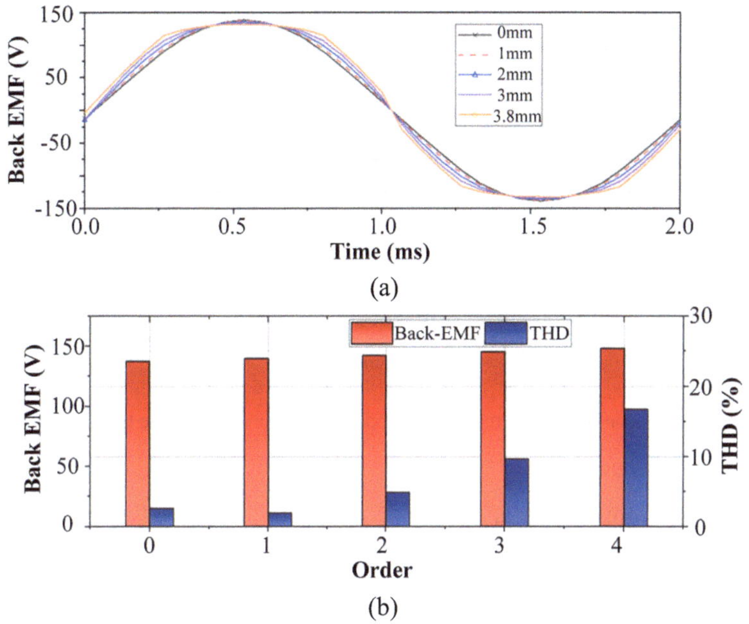

Fig. 2.62 Comparison of back-EMF with different PM edge thickness. **a** Back-EMF distribution. **b** Fundamental and total h armonic distortion of back-EMF

harmonics also increases. The proportion of fundamental component first increases more significantly, leading to a decrease in THD. However, as the PM shape approached surface-mount, the proportion of higher harmonics increases significantly, and THD gradually increased.

The comparison of back electromotive force (back-EMF) with different PM edge thickness as shown in Fig. 2.62. It can be clearly found that the fundamental component of conventinal surface-mounted PM machine is the highest, while this with a PM edge thickness of 0mm is the lowest. And the machine with PM edge thickness of 1mm has the lowest THD, which is consistent with the situation of the air gap magnetic field.

Figure 2.63 shows the torque characteristic with different edge thickness. It can be found that the average torque increases with the increase of edge thickness and the conventional one has the highest average output torque. Similar to flux density characteristics, torque ripple also decreases first and then increases. This result is consistent with flux density analysis, but due to the effect of the armature reaction, the minimum torque ripple occurs at an edge thickness of 2mm instead of the previous 1mm.

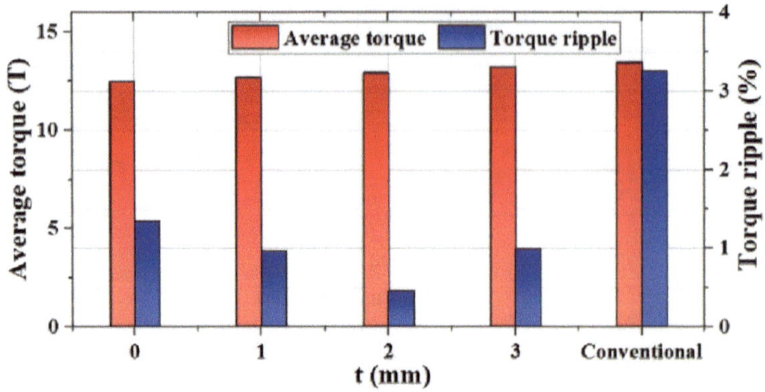

Fig. 2.63 Comparison of average torque and ripple with different PM edge thickness

2.4.2.2 The Comparison and Analysis of Electromagnetic Force

The circumferential path integration of tangential electromagnetic force of the machine forms torque, while radial electromagnetic force acts on the stator teeth, causing deformation and vibration, which is undesirable. Analyzing the radial and tangential electromagnetic forces of the machine plays a significant role in the study of vibration suppression.

Based on the superposition principle, the radial and tangential magnetic field in the air-gap can be expressed as

$$\begin{cases} B_r = B_x \cos\theta + B_y \sin\theta \\ B_t = B_x \sin\theta - B_y \cos\theta \end{cases} \tag{2.24}$$

where B_x, B_y is the x-axis and y-axis component of the air-gap magnetic field. According to the Maxwell stress tensor equation, the radial and tangential electromagnetic force can be approximated as

$$\begin{cases} p_r = \frac{1}{2\mu_0}\left(B_r^2 - B_t^2\right) \\ \\ p_t = \frac{1}{\mu_0}\left(B_r B_t\right) \end{cases} \tag{2.25}$$

It should be noted that for the convenience of subsequent vibration analysis, the influence of radial force will be mainly investigated.

The distribution of radial and tangential magnetic fields is shown in Fig. 2.64, the maximum value of the radial magnetic field is around 1. The maximum value of the tangential magnetic field is around 0.25, which is due to the effect of the armature magnetic field, resulting in a significant increase in this value compared to when it is unloaded.

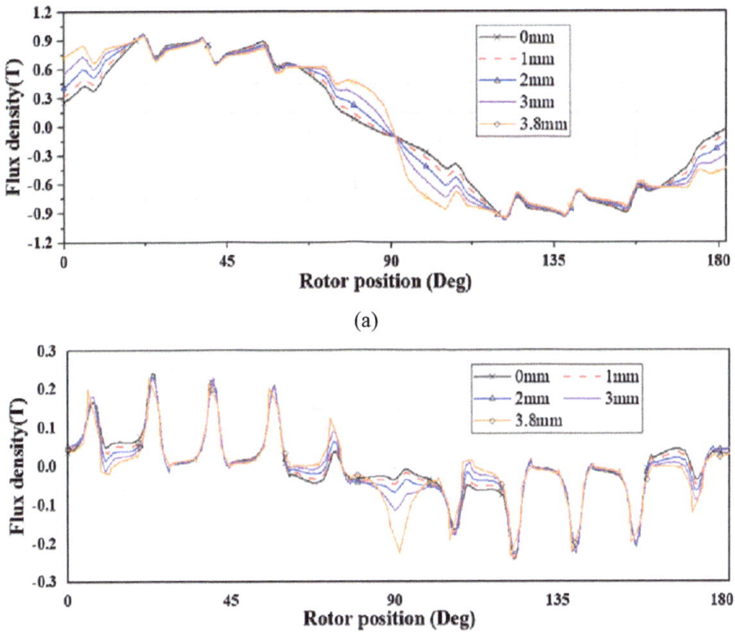

(a)

(b)

Fig. 2.64 Comparison of the radial and tangential magnet field with different edge thickness. **a** Radial magnet field. **b** Tangential magnet field

As displayed in Fig. 2.65a, the radial electromagnetic force in the machine air-gap is a wave function of time and space. The 2D fast Fourier transformation (FFT) of the radial force with different edge thickness is performed, and the results are shown in Fig. 2.65b. It can be clearly found that the radial electromagnetic force is mainly distributed in the *np* order in space, which is determined by the structure of the machine. Its positive and negative orders represent different machine rotation direction, and the components with larger amplitudes are concentrated in lower orders. In terms of tempory order, it is mainly distributed in even multiples of the machine's fundamental frequency. In comparison, it can be found that the 0th order component is always the largest. On the other hand, it can be found that the amplitude of the 0th order component increases with the increase of edge thickness, which is due to the gradual enhancement of the fundamental wave of air-gap flux density. Besides, the magnitudes of 4th-order radial force at 1000 Hz are all the largest, and the amplitude with the edge thickness of 0mm is the largest among all comparison machines due to the injection of the third-order harmonic. The distribution of the magnitudes of 12th-order radial force at 3000 Hz first decreases and then increases. In the high-frequency range, it is found that appropriate edge thickness like 2mm can reduce the amplitude of high-frequency radial electromagnetic force, which is helpful for weakening high-frequency vibration.

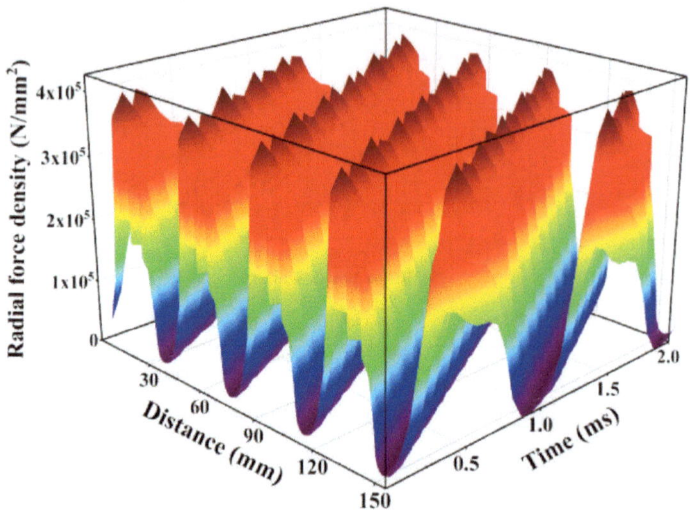

(a)

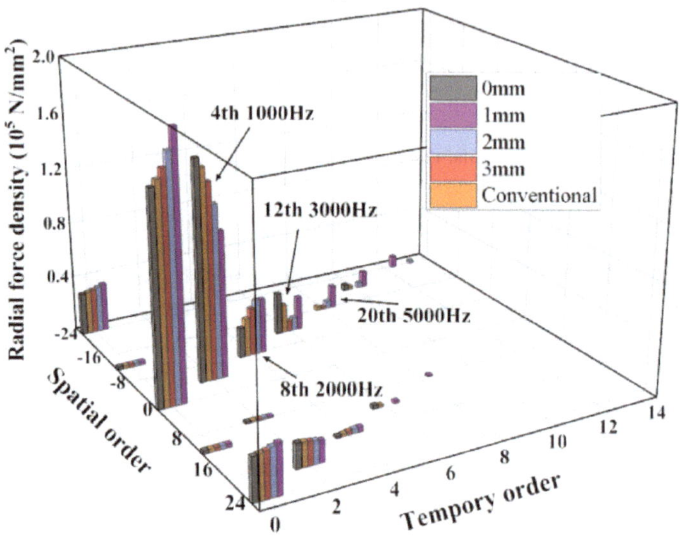

(b)

Fig. 2.65 Comparison of radial electromagnetic force. **a** Distribution of radial electromagnetic forces in space and time (for clarification, only results for t = 0 mm is shown). **b** Distribution of 2D FFT decomposition

2.4.2.3 The Comparison and Analysis of Vibration Response

When the spatial order of the radial electromagnetic force is equal to the modal order, and the frequency of the electromagnetic force at that order is also equal or similar to this modal frequency, it will generate severe vibration of the motor. Therefore, it is nessary to perform modal analysis on the machine before conducting vibration analysis. In order to reduce the impact of mechanical structure on vibration, the machine model does not consider the structures such as casing and end caps. The modal analysis adopts the free mode analysis method, and the low order modal shapes and frequencies obtained after ignoring the first six free modals are shown in Table 2.6 The axial modals are ignored because the main focus is on vibration caused by radial electromagnetic forces.

The simunlated vibration deformation and acceleration of machine with different edge thickness is disaplayed in Fig. 2.66a, it can be found that the peak value emerging at 1000 Hz. This result proves the correctness of electromagnetic force analysis. The 4th radial electromagnetic force at 1000 Hz is the main source of the peak

Table 2.6 Modal parameters from FEA simulation

Modal shape	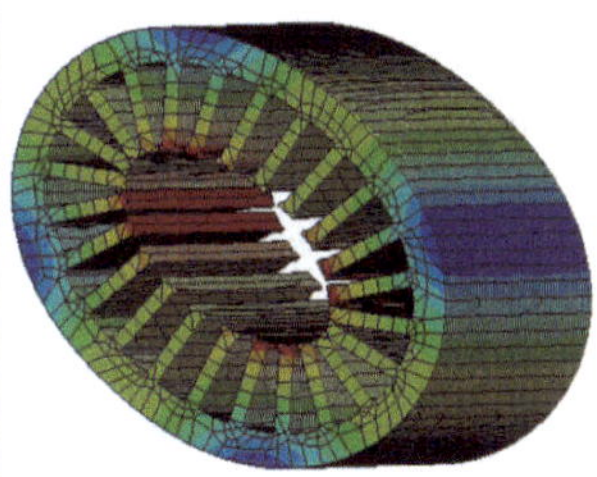	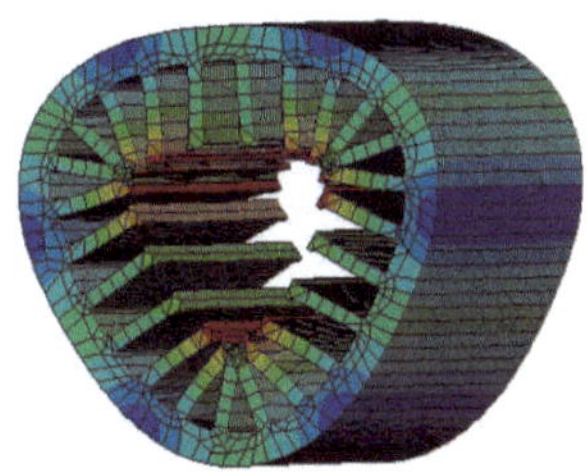
Order	2	3
Frequency	1468.2Hz	3493.3Hz
Modal shape	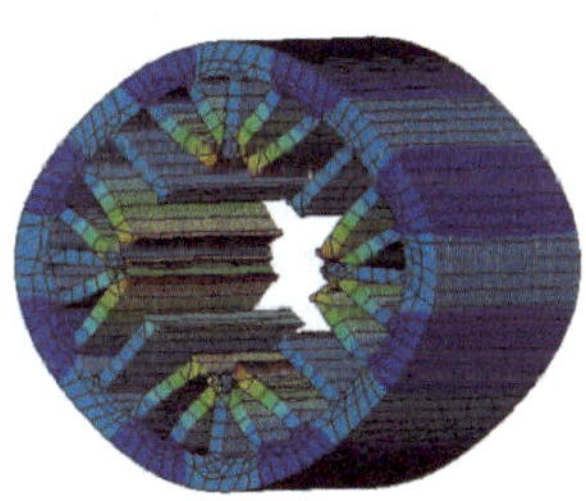	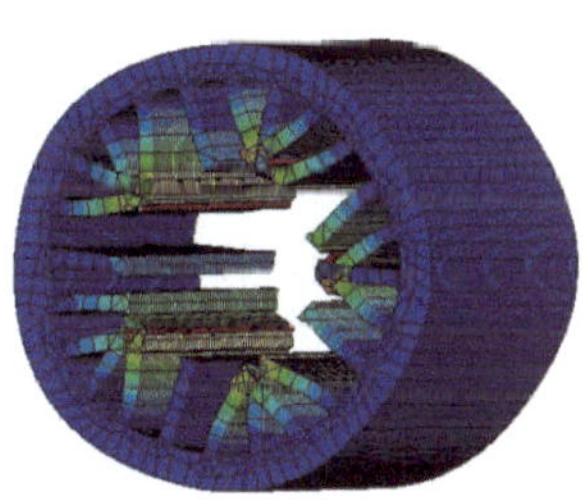
Order	4	5
Frequency	4892.6Hz	5436.2Hz

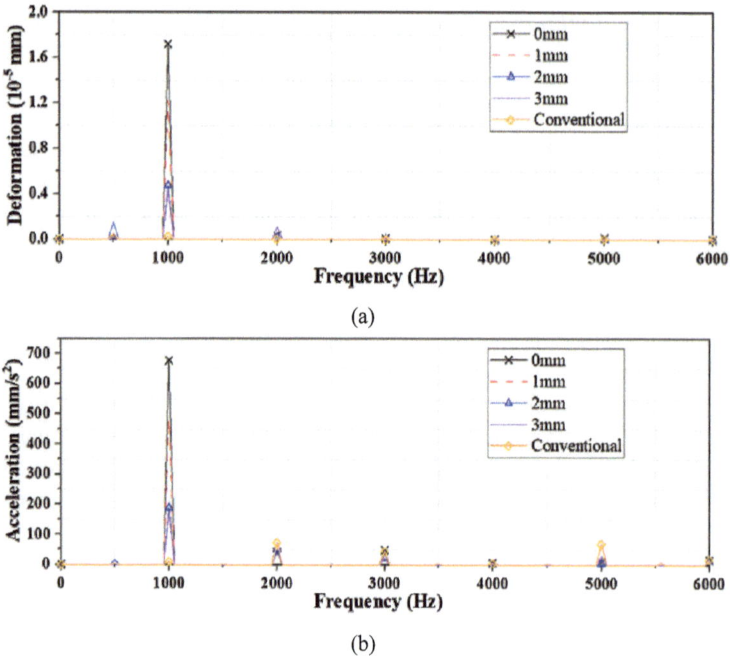

Fig. 2.66 Comparion of the frequency response. **a** Deformation at different frequencies. **b** Acceleration at different frequencies

vibration of this machine. Consistent with the distribution of radial electromagnetic force, the vibration at 1000 Hz also decreases with the increase of edge thickness, which indicates that the 3th harmonic PM will damage the vibration performance of the machine. The deformation at 2000 Hz is also relatively large, which is caused by the larger amplitude of the 8th radial electromagnetic force, so the vibration acceleration at this frequency is also larger. On the other hand, the vibration acceleration at 5000 Hz is also a potential danger, the 20th electromagnetic force is its main source. The amplitude of the 5th radial electromagnetic force is not large and can not generate significant vibration deformation, but it can generate significant vibration acceleration due to the high frequency of 5000 Hz. Overall, the main source of vibration for this machine is the 4th radial electromagnetic force. The vibration response could be improved by eliminating or suppressing this order of electromagnetic force. For this model, choosing 2mm as the edge thickness would be a perfect answer. The increase in vibration deformation and acceleration of the 3rd harmonic PM under this thickness is acceptable compared to the conventional surface-mounted version.

2.5 Conclusion

In this chapter, the quasi-cylindrical PM poles and a hybrid-irregular pole pattern consisting of quasi-T-shaped magnets and surface-mounted magnets are introduced to improve the torque output and depress the torque ripple simultaneously. The fundamental components of airgap flux density and the back EMF of the quasi-cylindrical PM poles can be increased, and thus the torque output can be improved significantly. It can increase the average torque by 9.29% and 9.04%, with respect to the eccentric and radial pole patterns, respectively. The high-order flux terms can be reduced effectively, and thus the torque ripple can be depressed well. Compared with the conventional eccentric and radial pole patterns, the quasi-cylindrical pole pattern reduces the torque ripple by 17.76% and 6.02%, respectively. In addition, due to the uniform magnetomotive force along the circumference in quasi-cylindrical PM pole pattern, the anti-demagnetization capability is better than the conventional eccentric and radial pole patterns. And the quasi-T-shaped magnets not only helps to improve flux-focusing effect, but also are conducive to reduce high-order air-gap magnetic harmonics and hence achieving smooth torque output. Compared with the conventional Spoke-type PM machine, the proposed hybrid-irregular pole pattern has several advantages as follow. Firstly, the average torque of the electric machine with hybrid-irregular pole pattern is increased by 15.54% form 13.58Nm to 15.69Nm, which verifies the stronger flux-focusing effect of the proposed design compared with Spoke-type design. Secondly, the torque ripple of the electric machine with the proposed pole pattern is reduced by 74.42% compared to the conventional one. The higher power density (2.73kW/kg) and higher operating efficiency (97.78%) than those of the conventional Spoke-type design can be achieved by the employment of the hybrid-irregular pole pattern. Menawhile, a 24-slots/4-poles PM machine with third-order harmoncis PMs is designed and the shape functions of third-order harmonic PM with different edge thickness are presented to investigate the effect of the PM edge thickness on performance of machine. The comparison result of the no-load magnetic field shows that as the edge thickness increased, the sinusoidal characteristics of the air-gap flux density distribution gradually deteriorated, but the fundamental component gra-dually increased, and the THD showed a trend of first de-creasing and then increasing, reaching its minimum at 1mm. This reveals the role of edge thickness in third-order harmonic PM, which controls the proportion of third-order harmonics in the conventional surface-mounted type. The comparison results of the output performance of PM machine are consistent with the analysis of the magnetic field, but the machine with a 2mm PM edge thickness has the optimal output performance due to the effect of armature magnetic field. Moreover, the PM edge thickness can increase the amplitude of low-order radial electromagnetic force, thereby deteriorating the low-frequency vibration of machines. But in high-frequency range, the appropriate PM edge thickness can weaken the electromagnetic amplitude and suppress the high-frequency vibration, which is significant to supress the vibration and noise of PM machine. Relatively, setting the PM edge thickness at half or one-third of the conventional surface-mount thickness will help reduce torque ripple and minimize damage to vibration performance, while also increasing mechanical reliability and machining feasibility.

References

1. Z. Q. Zhu, L. J. Wu, and Z. P. Xia, "An accurate subdomain model for magnetic field computation in slotted surface-mounted permanent-magnet machines, IEEE Trans. Magn., vol. 46, no. 4, pp. 1100–1115, Apr. 2010
2. Y. Zeng, M. Cheng, G. Liu, and W. Zhao, "Effects of magnet shape on torque capability of surface-mounted permanent magnet machine for servo applications," IEEE Trans. Ind. Electron., vol. 67, no. 4, pp. 2977–2990, Apr. 2020.
3. G. Liu, Y. Zeng, W. Zhao, and J. Ji, "Permanent magnet shape using analytical feedback function for torque improvement," IEEE Trans. Ind. Electron., vol. 65, no. 6, pp. 4619–4630, 2018.
4. G. Heins, M. Thiele, and T. Brown, "Accurate torque ripple measurement for PMSM," IEEE Trans. Instrum. Meas., vol. 60, no. 12, pp. 3868–3874, Dec. 2011.
5. H. Y. Sun and K. Wang, "Effect of third harmonic flux density on cogging torque in surface-mounted permanent magnet machines," IEEE Trans. Ind. Electron., vol. 66, no. 8, pp. 6150–6158, 2019.
6. X. Ge, Z. Q. Zhu, G. Kemp, D. Moule, and C. Williams, Optimal step-skew methods for cogging torque reduction accounting for threedimensional effect of interior permanent magnet machines, IEEE Trans. Energy Comers., vol. 32, no. 1, pp. 222–232, 2017.
7. Z Shi et al., "Torque analysis and dynamic performance improvement of a PMSM for EVS by skew angle optimization, IEEE Trans. ApplSupercond, vol. 29, no. 2, pp. 1–5, 2019.
8. C. Peng, D. Wang, Z Feng, and B. Wang, "A new segmented rotor to mitigate torque ripple and electromagnetic vibration of interior permanent magnet machine, IEEE Trans. Ind. Electron., vol. 69. no. 2, pp. 1367–1377, Feb. 2022.
9. SS R. Bonthu, M. T. B. Tarek, and S. Choi, "Optimal torque ripple reduction technique for outer rotor permanent magnet synchronous reluctancemotors, IEEE Trans. Energy Convers., vol. 33, no. 3, pp. 1184–1192, Mar. 2018.
10. N. Bianchi and S. Bolognani, "Design techniques for reducing the cogging torque in surface-mounted PM motors," IEEE Trans. Ind. Appl., vol. 38, no. 5, pp. 1259–1265, 2002.
11. Y. Yokoi and T. Higuchi, "Stator design of alternate slot winding fo reducing torque pulsation with magnet designs in surface-mounted permanent magnet motors," IEEE Trans. Magn., vol. 51, no. 6, Jun. 2015.
12. L. Wu and Z. Q. Zhu, "Analytical modeling of surface-mounted PM machines accounting for magnet shaping and varied magnet property dis-tribution, " IEEE Trans. Magn., vol. 50, no. 7. Jul. 2014.
13. S. Ruangsinchaiwanich, Z. Zhu, and D. Howe, "Influence of magnet shape on cogging torque and back-emf waveform in permanent magnet machines," Proc. Int. Conf. Elect. Mach. Syst., 2005, pp. 284–289.
14. Y. Li, J. Zou, and Y. Lu, "Optimum design of magnet shape in permanent-magnet synchronous motors," IEEE Trans. Magn., vol. 39, no. 6, pp. 3523–3526, 2003.
15. W. Zhao, T. A. Lipo, and B.-I. Kwon, "Material-efficient permanent-magnet shape for torque pulsation minimization in SPM motors for automotive applications," IEEE Trans. Ind. Electron., vol. 61, no. 10, pp. 5779–5787, 2014.
16. P. Upadhayay and K. R. Rajagopal, "Torque ripple reduction using magnet pole shaping in a surface mounted permanent magnet BLDC motor," Proc. Int. Conf. Renewable Energy Res. Appl., 2013, pp. 516–521.

17. S. Asgari and M. Mirsalim, “A novel dual-stator radial-flux machine with diametrically magnetized cylindrical permanent magnets," IEEE Trans. Ind. Electron., vol. 66, no. 5, pp. 3605–3614, 2019.
18. Y. Zhou, H. Li, G. Meng, S. Zhou, and Q. Cao, “Analytical calculation of magnetic field and cogging torque in surface-mounted permanent-magnet machines accounting for any eccentric rotor shape," IEEE Trans. Ind. Electron., vol. 62, no. 6, pp. 3438–3447, Jun. 2015.
19. K. Wang, “Effects of harmonics into magnet shape and current of dual three-phase permanent magnet machine on output torque capability," IEEE Trans. Ind. Electron., vol. 65, no. 11, pp. 8758–8767, Nov. 2018.
20. K. Wang, Z. Y. Gu, C. Liu, and Z. Q. Zhu, “Design and analysis of a five-phase SPM machine considering third harmonic current injection," IEEE Trans. Energy Convers., vol. 33, no. 3, pp. 1108–1117, 2018.
21. Z. S. Du and T. A. Lipo, “High torque density and low torque ripple shaped-magnet machines using sinusoidal plus third harmonic shaped magnets," IEEE Trans. Ind. Appl., vol. 55, no. 3, pp. 2601–2610, 2019.
22. L. F. Zhang, K. Wang, H. Y. Sun, and S. Zhu, “Multi-phase PM machines with halbach array considering third harmonic flux density," IEEE Trans. Ind. Electron., vol. 66, no. 12, pp. 9184–9193, 2019.
23. K. Wang, Z. Q. Zhu, and G. Ombach, “Torque enhancement of surface- mounted permanent magnet machine using third-order harmonic," IEEE Trans. Magn., vol. 50, no. 3, pp. 104–113, 2014.
24. P. Hu et al., “The modified model of third-harmonic shaping for a surface-mounted permanent-magnet synchronous motor under parallel magnetization," IEEE Trans. Ind. Appl., vol. 56, no. 5, pp. 4847–4856, 2020.
25. Kim, Sung-An et al., “Optimal Rotor Shape Design of 3-step Skew Spoke Type BLAC Motor to Reducing Cogging Torque," 2016.
26. P. Liang, F. Chai, Y. Yu and L. Chen, “Analytical Model of a Spoke-Type Permanent Magnet Synchronous In-Wheel Motor With Trapezoid Magnet Accounting for Tooth Saturation," IEEE Transactions on Industrial Electronics, vol. 66, no. 2, pp. 1162–1171, 2019.
27. X. Zhu, W. Wu, L. Quan, Z. Xiang and W. Gu, “Design and Multi-Objective Stratified Optimization of a Less-Rare-Earth Hybrid Permanent Magnets Motor With High Torque Density and Low Cost," IEEE Transactions on Energy Conversion, vol. 34, no. 3, pp. 1178–1189, 2019.
28. B. Xia and P. C. -K. Luk, “Analytical model of open-circuit characteristics of two-layer spoke-type ferrite interior permanent magnet machines," 2017 IEEE International Electric Machines and Drives Conference (IEMDC), Miami, FL, USA, 2017, pp. 1–8.
29. Y. Mao, W. Zhao, S. Zhu, Q. Chen and J. Ji, “Vibration Investigation of Spoke-Type PM Machine With Asymmetric Rotor Considering Modulation Effect of Stator Teeth," IEEE Transactions on Industrial Electronics, vol. 68, no. 10, pp. 9092–9103, 2021.
30. X. Ge, Z. Q. Zhu, J. Li and J. Chen, “A Spoke-Type IPM Machine With Novel Alternate Airspace Barriers and Reduction of Unipolar Leakage Flux by Step-Staggered Rotor," IEEE Transactions on Industry Applications, vol. 52, no. 6, pp. 4789–4797, Nov.-Dec. 2016.
31. Y. Xiao, Z. Q. Zhu, G. W. Jewell, J. Chen, D. Wu and L. Gong, “A Novel Spoke-Type Asymmetric Rotor Interior Permanent Magnet Machine," IEEE Transactions on Industry Applications, vol. 57, no. 5, pp. 4840–4851, Sept.-Oct. 2021.
32. J. Park, R. Tsunata, M. Takemoto, K. Orikawa and S. Ogasawara, “Investigation of Dy-Free Hybrid PM Motor Based on Spoke-Type Rotor for Automotive Applications," 2021 IEEE International Electric Machines and Drives Conference (IEMDC), Hartford, CT, USA, 2021, pp. 1–8.
33. Y. Liu, L. Li, Q. Gao, J. Cao and Z. Sun, “An Analytical Model and Optimization of a Novel Hybrid Rotor Machine for High Torque Density," IEEE Transactions on Energy Conversion, vol. 36, no. 1, pp. 230–241, 2021.
34. S. Asgari and M. Mirsalim, “A novel dual-stator radial-flux machine with diametrically magnetized cylindrical permanent magnets," IEEE Trans. Ind. Electron., vol. 66, no. 5, pp. 3605–3614, 2019.

35. S. Huang, J. Luo, F. Leonardi, and T. Lipo, "A general approach to sizing and power density equations for comparison of electrical machines," IEEE Trans. Ind. Appl., vol. 34, no. 1, pp. 92–97, 1998.
36. H. Chen, N. A. O. Demerdash, A. M. EL-Refaie, Y. Guo, W. Hua and C. H. T. Lee, "Investigation of a 3D-Magnetic Flux PMSM With High Torque Density for Electric Vehicles," IEEE Transactions on Energy Conversion, vol. 37, no. 2, pp. 1442–1454, June 2022.
37. P. Xiang, L. Yan, Y. Guo, X. He, C. Gerada and I. -M. Chen, "A Concentrated-Flux-Type PM Machine With Irregular Magnets and Iron Poles," IEEE/ASME Transactions on Mechatronics, https://doi.org/10.1109/TMECH.2023.3293505.
38. J. Ahn, et al. "Correlation between rotor vibration and mechanical stress in ultra-high-speed permanent magnet synchronous motors," IEEE Transactions on Magnetics, vol. 53, no. 11, pp. 1–6, 2017.
39. L. Jing, et al. "Investigation of a surface PM machine with segmented-eccentric magnet poles," IEEE Transactions on Applied Superconductivity, vol. 28, no. 3, pp. 1–5, 2018.
40. S. T. Boroujeni et al. "Analytical modeling of no-load eccentric slotted surface-mounted PM machines: Cogging torque and radial force," IEEE Transactions on Magnetics, vol. 53, no. 12, pp. 1–8, 2017.
41. Y. Zhou et al. "Analytical calculation of magnetic field and cogging torque in surface-mounted permanent-magnet machines accounting for any eccentric rotor shape." IEEE transactions on industrial electronics, vol. 62, no. 6, pp. 3438–3447, 2014.
42. T. Liu et al. "Effects of eccentric magnet on high-frequency vibroacoustic performance in integral-slot SPM machines." IEEE Transactions on Energy Conversion, vol. 36, no. 3, pp. 2393–2403, 2021.
43. H. Lin et al. "Influence of magnet shape on torque behavior in surface-mounted permanent magnet motors." 2014 17th International Conference on Electrical Machines and Systems (ICEMS). IEEE, 2014.
44. Y. Li, J. Zou, and Y. Lu. "Optimum design of magnet shape in permanent-magnet synchronous motors." IEEE transactions on magnetics, vol. 39, no. 6, pp. 3523–3526, 2003.
45. K. Wang, Z. Q. Zhu, and G. Ombach. "Torque enhancement of surface-mounted permanent magnet machine using third-order harmonic." IEEE Transactions on Magnetics, vol. 50, no. 3, pp. 104–113, 2013.
46. G. Liu et al. "Permanent magnet shape using analytical feedback function for torque improvement." IEEE Transactions on Industrial Electronics, vol. 65, no. 6, pp. 4619–4630, 2017.
47. J. Ji, et al. "Vibration reduction and torque improvement of integral-slot SPM machines using PM harmonic injection." Chinese Journal of Electrical Engineering, 2023.

Chapter 3
Pole Pattern Topology for High Torque and Lightweight Rotor

Abstract This chapter introduces two methods for improving output torque while reducing rotor mass. The novel concentrated-flux-type PM machine with irregular magnets and iron poles (CPMIP) is considered as a topology structure that can effectively achieve this goal. The configuration has following advantages. First, the employment of quasi-trapezoid magnets with opposite magnetization direction and large cross-section helps to enhance the flux-focusing effect and thus the torque generation. Second, the quasi-triangular iron poles can improve the magnetic self-shielding effect well, and thus the rotor yoke can be removed completely and the rotor mass could be reduced significantly. Third, due to the reduction of PM volume and the improvement of torque output, the PM utilization ratio of electrical machine can be increased greatly. On the other hand, the working principles of linear machines and rotary machines are the same, so consistent conclusions that can be extended to the field of rotary machine design can be obtained by studying certain linear machines. The novel compound Halbach array consisting of the outer layer of quasi-Halbach array and inner layer of axial magnetized pattern is considered an good candidate for reducing rotor mass while improving output performance. This magnetic pattern can completely removed the high-density back iron existed in conventional linear PM machines due to the much improvement of self-shielding effect. Thus, the mover mass can be greatly reduced and hence high dynamic performance can be obtain.

3.1 Introduction

Due to the merits of high working efficiency and large power density, permanent-magnet (PM) machines have been extensively employed in many industrial applications, aerospace facilities and electric vehicles [1–3]. The applications of electrical machines in electric aircrafts mainly concern three aspects, i.e., torque output, rotor mass and magnet volume. The high torque output helps to improve the pull force of electrical vehicle, the low rotor mass aids in high dynamic response, and small rare-earth magnet volume can reduce the eddy loss and the manufacturing cost greatly. However, it is generally challenging to improve these three aspects, simultaneously.

P. Xiang et al., *Rotor Pole Pattern Topology Technology for Permanent Magnet Electric Machine*, https://doi.org/10.1007/978-981-95-2018-3_3

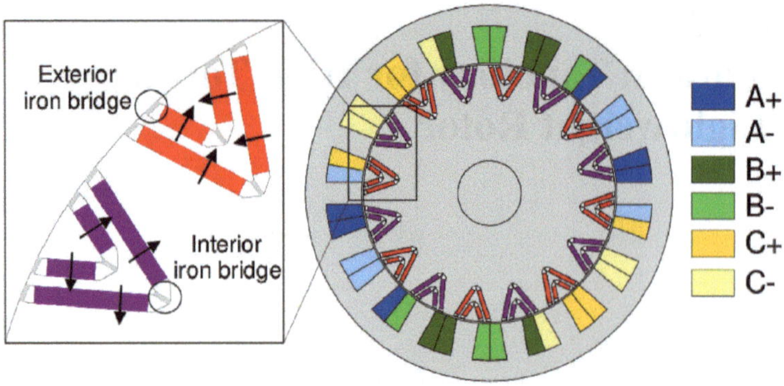

(a)

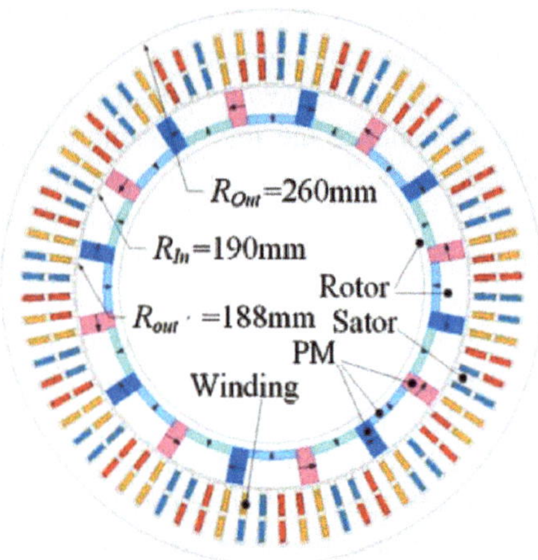

(b)

Fig. 3.1 Research model. **a** Studied motor structure in [4]. **b** Cross section of the proposed PM machine in [5]

To improve torque output, interior PM (IPM) machines are usually employed, due to their additional reluctance torque [4, 5]. The specific rotor profile of this research is shown in Fig. 3.1. Among various IPM machines, spoke-type machines are very popular due to the strong flux-focusing effect of two adjacent PMs except for the reluctance torque [6], which could further enhance the airgap flux density and hence output capability. As shown in Fig. 3.2, Xu et al. [7] employed a spoke-type pole pattern, i.e., rectangular magnet and trapezoid iron of two adjacent PMs, in vernier machines for electric vehicles to improve system torque output and power factor. A novel alternate airspace barriers were proposed to reduce unipolar leakage of the spoke-type pole pattern and thus enhance torque output and bearing life [6].

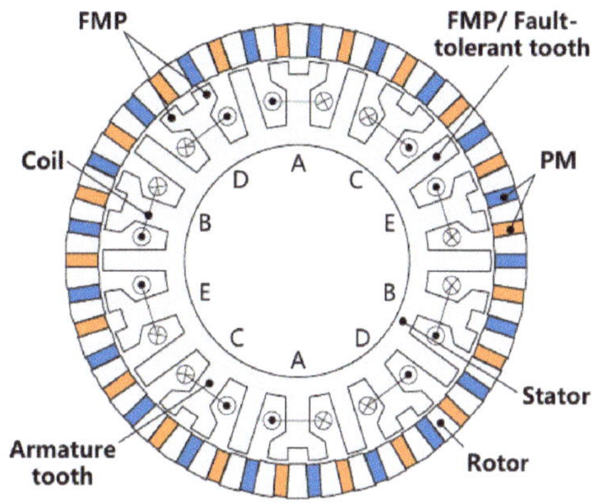

Fig. 3.2 Structure of the STPMV machine in [7]

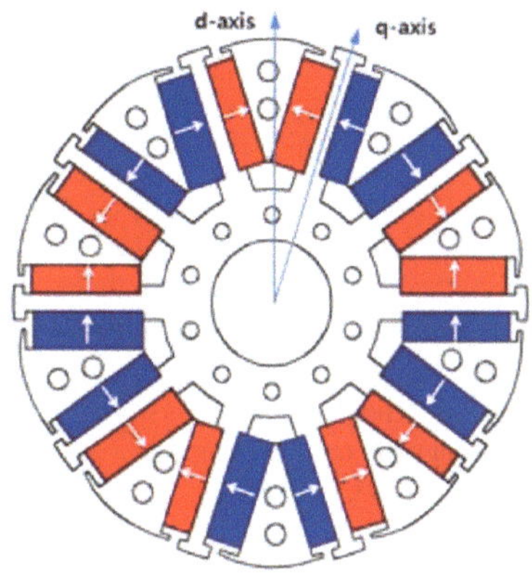

Fig. 3.3 Rotor configurations designed for inner bridged spoke-type FMM in [8]

As shown in Fig. 3.3, Kim et al. [8] proposed a similar spoke-type machine but with unequal thickness rectangular magnets in magnetization direction, which improves its anti-demagnetization capability along with some distortion in airgap flux density waveform. In fact, the airgap flux distortion generally exists in spoke-type machines and thus may result in torque ripple [9], as shown in Fig. 3.4. In [10], to increase flux density and hence torque output, pushing assistant magnets were inserted into the spoke-type machine, as shown in Fig. 3.5. In [11], a spoke-type machine with L-shaped magnets was investigated, as shown in Fig. 3.6, to improve the flux-focusing effect and thus airgap flux density in some extend. Overall, spoke-type machines are promising candidates to improve torque output. However, it can be found that the magnets of almost all spoke-type machines are rectangular shape, therefore, to achieve strong airgap flux, large cross-section rectangular magnets in magnetized direction are usually employed, which results in thick rotor and hence rotor mass contributing to poor dynamic performance and servo control characteristics [12]. In fact, to obtain large reluctance torque, large area of iron between magnets are

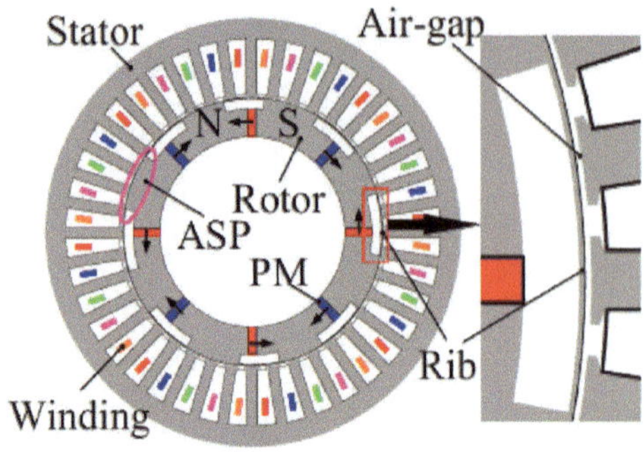

Fig. 3.4 Cross sections of the spoke-type PM machine with air-gap barriers in [9]

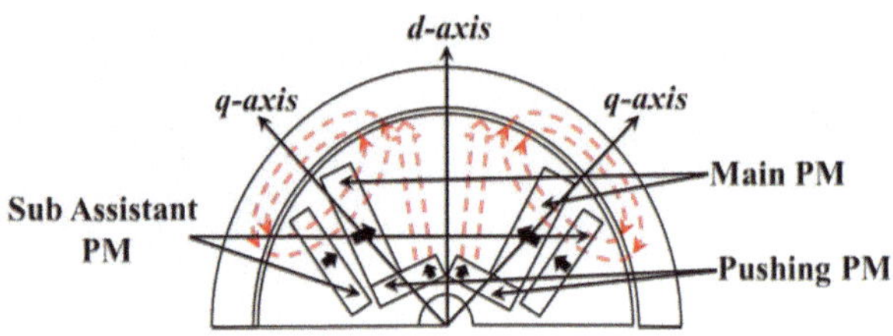

Fig. 3.5 Magnet flux characteristic of proposed wing shaped spoke-type PM rotor in [10]

Fig. 3.6 Rotor core and non-magnetic material shape of the L-type CFFM [11]

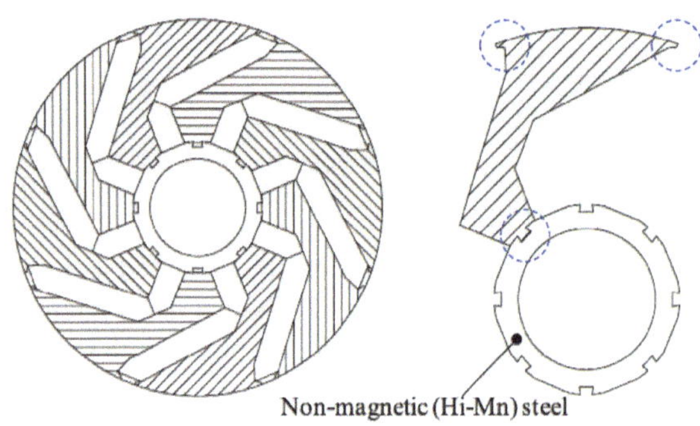

generally used, which may further increase rotor mass. In addition, their unipolar flux leakage also constrains the improvement of torque output [13]. And some methods of overcoming the drawback of flux leakage usually cause deterioration of rotor mechanical strength more or less.

To reduce rotor mass and hence achieve high dynamic performance, Halbach array along with its variants is an impressive magnet array due to its self-shielding effect [14]. For example, the quasi-Halbach array with radial and tangential magnetized PMs can weaken magnetic field in iron yoke and thus reduce the yoke mass and inertia effectively, because hollow structures can be employed for the rotor

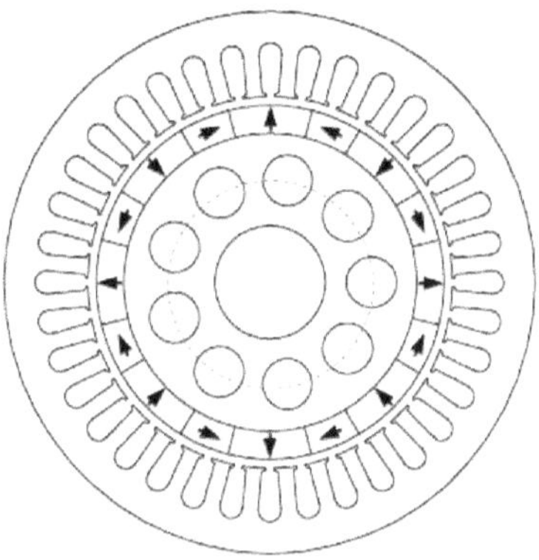

Fig. 3.7 Basic structure of the compact hydrostatic-driven electric generator in [14]

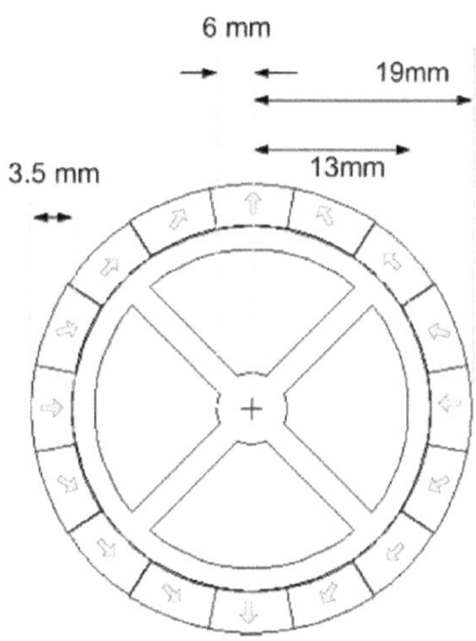

Fig. 3.8 Dimensions of Halbach-array-based rotor and the magnetization direction of the segments [15]

design [15], as shown in Figs. 3.7 and 3.8. Zhu et al. [16] studied various segmented Halbach arrays, as shown in Fig. 3.9, to approximate ideal Halbach array, in which the magnet of one pole is divided into multiple segments with different magnetization directions. As shown in Fig. 3.10, Zhang et al. [17] developed a variant of Halbach array with trapezoidal magnets for electromagnetic drive system. The air-gap flux density distribution and the self-shielding effect can be adjusted by the central angle of the magnet. However, generally the high-density rotor yoke can not be removed completely for the reduction of magnetic resistance due to the incomplete self-shielding effect of Halbach array, which unavoidably compromises system dynamics.

To achieve high PM utilization ratio, i.e., less PM mass and volume while maintaining torque output, consequent-pole PM (CPM) surface-mounted machine is one popular solution [18, 19]. The feature of this machine is that there are only N- or S-pole for reducing the amount of magnets, as shown in Figs. 3.11, 3.12 and 3.13.

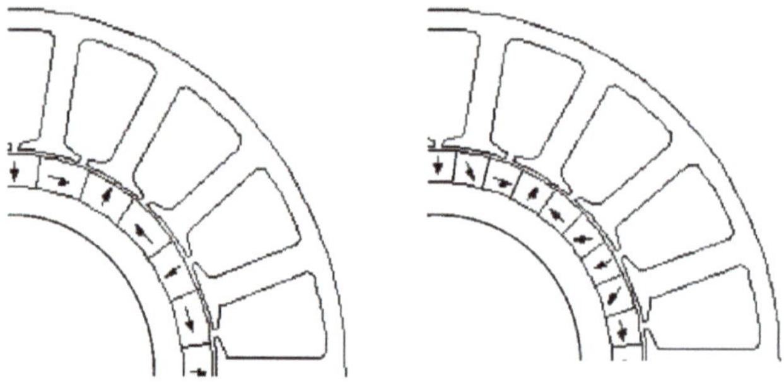

Fig. 3.9 Segmented Halbach magnet machine [16]

Fig. 3.10 A variant of Halbach array with trapezoidal magnets in [17]

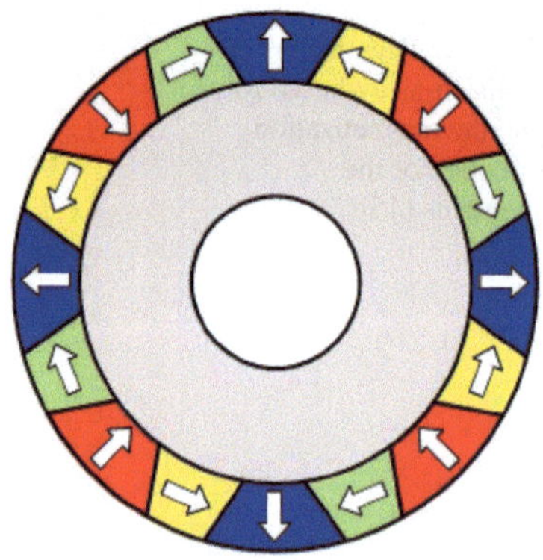

Fig. 3.11 Cross-sectional view of 12-/11-pole PS-FRPM machines with existing SPM and proposed CPM inner stators in [21]

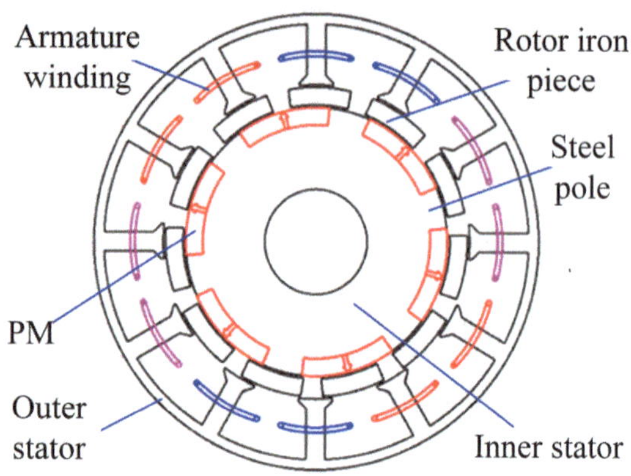

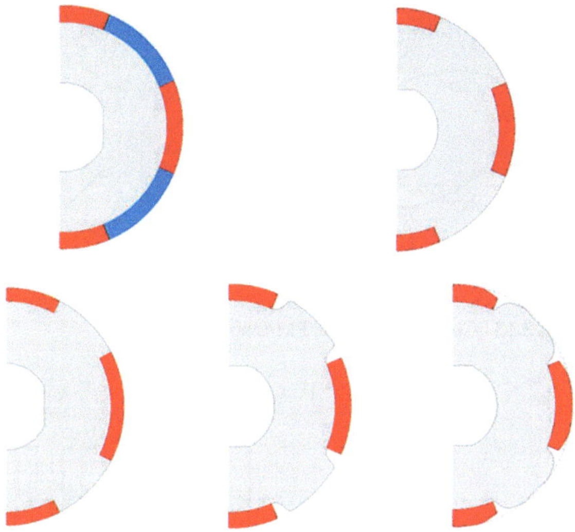

Fig. 3.12 Optimized consequent-pole models in [18]

Chung et al. [20] proposed a doubly salient PM linear machine with consequent-pole structure, as shown in Fig. 3.14, and only a half amount of PMs are used but with the reduction of thrust density. Similar results can be found in other CPM machine designs such as rotor-PM and stator-PM machines [21–23]. However, the average torque output of CPM machines are generally less than that of their SPM counterparts with alternatively magnetized magnets. To increase the torque output, the optimal PM-arc ratio, i.e., the ratio of the airgap length to the PM thickness, of the CPM rotor was analyzed via finite element method and analytically derived in [24, 25]. Besides, as shown in Fig. 3.15, a dovetailed consequent-pole rotor were proposed to reduce airgap reluctance and hence to improve the output torque to some extent [26]. However, in general, permanent magnets are replaced with iron poles in the design of CPM machines, which may reduce the magnetomotive force (MMF) and influence the torque generation more or less. Besides, only single magnet pole (N- or S-pole) would cause asymmetric distribution of airgap flux, which results in serious torque ripple, unbalanced magnetic force and unipolar leakage in the machine end regions [27]. In addition, high mass density ferromagnetic yoke is usually for the rotor design to decrease magnetic saturation, which in turn may influence the system dynamics and torque density as well.

Overall, the spoke-type machines have large torque density, the SPM machines with Halbach array have low rotor mass, and the CPM surface-mounted machines have high PM utilization ratio. However, there are few literatures focus on improving these three aspects, simultaneously. Therefore, one of the objective of this chapter is to introduce a novel concentrated-flux-type PM machine with irregular magnets and iron poles to improve torque output and PM utilization ratio and reduce rotor mass, simultaneously. It offers several advantages as follows. First, the employment of quasi-triangular iron poles and alternatively tangential magnetization trapezoid

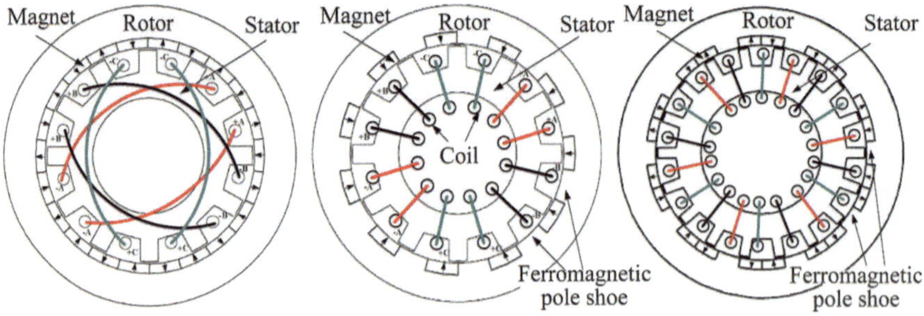

Fig. 3.13 Cross section of VPM topologies in [24]

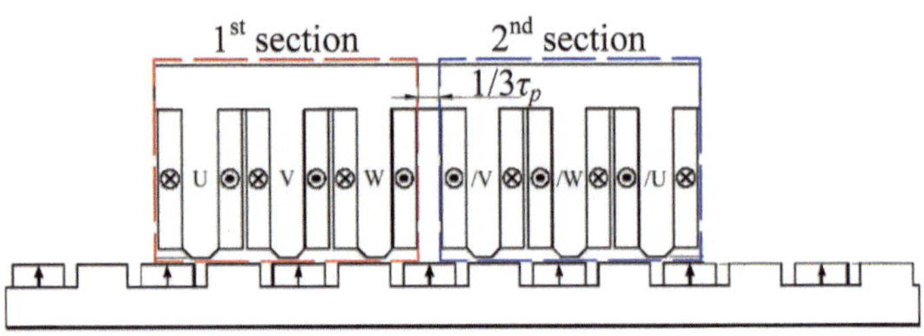

Fig. 3.14 Doubly salient PM linear machine with consequent-pole structure in [20]

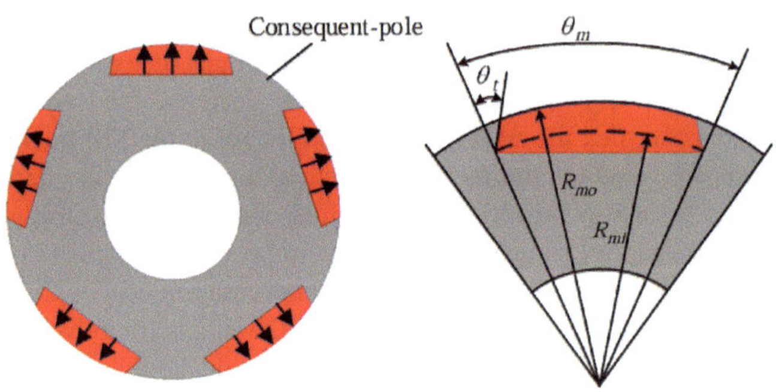

Fig. 3.15 Dovetailed consequent-pole rotor in [26]

magnets helps to enhance the flux-focusing effect, and thus the flux density in the airgap as well as the torque output can be increased effectively. Second, the configuration can improve the self-shielding effect of pole pattern significantly. Thus, there is no flux in the rotor yoke, and the high-density ferromagnetic material in the yoke can be replaced with low-density materials such as aluminum or polymers. This replacement can reduce the rotor mass, increase the system dynamics greatly. Third, compared with conventional SPM counterpart, due to the reduction of PM volume about 36.5% and the improvement of torque output, the PM utilization ratio of the proposed CPMIP machine can be improved significantly. Compared with conventional radial pattern and Halbach array, the CPMIP-2 magnet pattern has 71% and 64% higher PM utilization ratio, respectively. Besides, compared with conventional CPM pattern, the CPMIP-2 pattern improves the average torque by 29.2%, PM utilization ratio by 12.3% and reduces the rotor yoke mass by 63.53%.

On the other hand, in the field of linear machines, there has been extensive research on improving output characteristics and dynamic response. With the merits of high efficiency and simple structure, linear permanent magnet oscillating actuator, which can directly produce short-stroke reciprocating linear motion and thrust output without any rotation-to-translation conversion mechanisms, is widely used in industrial applications such as compressors and linear pumps [30, 31]. The high-performance PM machines, which are used in many industrial applications, attract much attention because of the utilization of rare-earth PMs with high magnetic energy product [32–34]. In recent years, high power density and high-frequency actuator is becoming a trend and has a growing demand, especially in aerospace applications, such as linear electro-hydrostatic actuator [35]. Thus, many researchers strive to improve its dynamic performance through increasing the force output or decreasing mover mass, i.e., low inertia [36].

To increase force output, raising input current under reasonable thermal design is a feasible method, however, it may cause high temperature more or less. Thus, some studies focus on changing pole pattern to enhance flux field of airgap [37]. Radial and axial magnetization arrays are two typical patterns [38, 39]. In generally, due to the flux focusing effect of axially magnetized pattern, it exhibits higher force output than that of radially magnetized pattern, as shown in Fig. 3.16. However, the axially magnetized pattern also has higher ripple [40]. To improve the force output of radial one, a square-shaped cross section tubular linear machine, as shown in Fig. 3.17, is proposed [41]. Baker et al. [42] develops an air-cored stator tubular linear generator with axially magnetized pattern for marine energy converters, as shown in Fig. 3.18, which attributes to high power factor. And the design of air-cored stator is beneficial to the reduction of force ripple. Huang et al. [43] employs an auxiliary end teeth design in the axially magnetized pattern, as shown in Fig. 3.19, and hence the force output is increased in a certain degree. In [44], an improved convex pole with same axial magnetization is proposed to increase thrust output and reduce the usage of PMs simultaneously, as shown in Fig. 3.20.

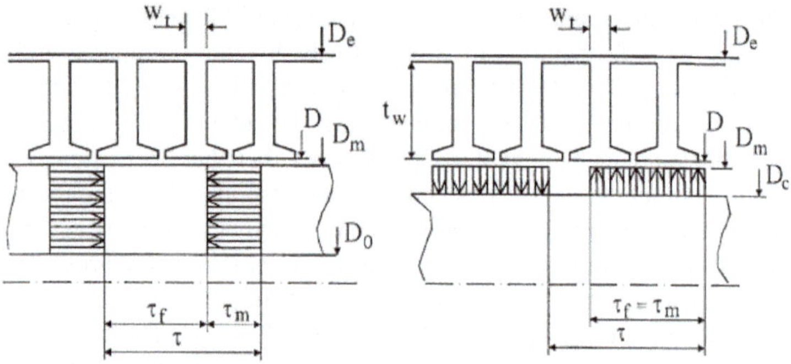

Fig. 3.16 Sketch of TL-IPM motor and TL-SPM motor in [40]

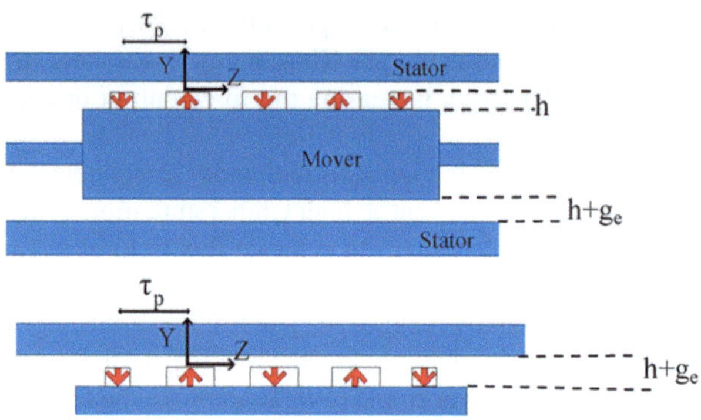

Fig. 3.17 TPLM with the upper-side mover PMs and its coordinate axes in [41]

Fig. 3.18 The research model in [42]

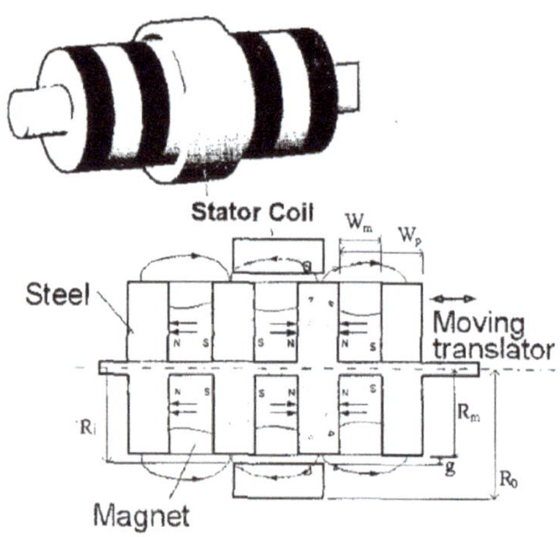

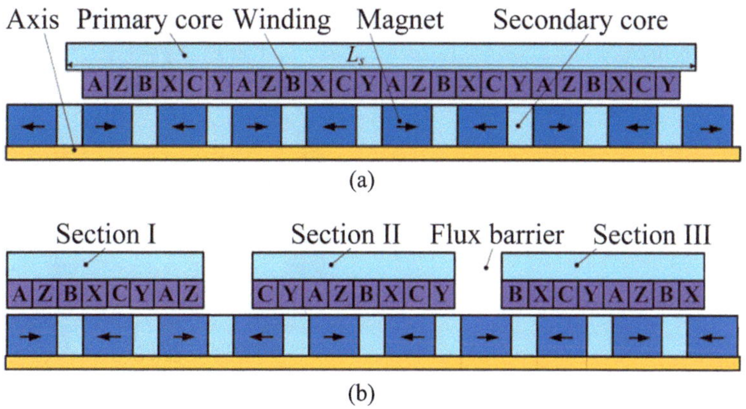

Fig. 3.19 Modular slotless TPMLSM in [43]. **a** Primary integrated core. **b** Segmented core

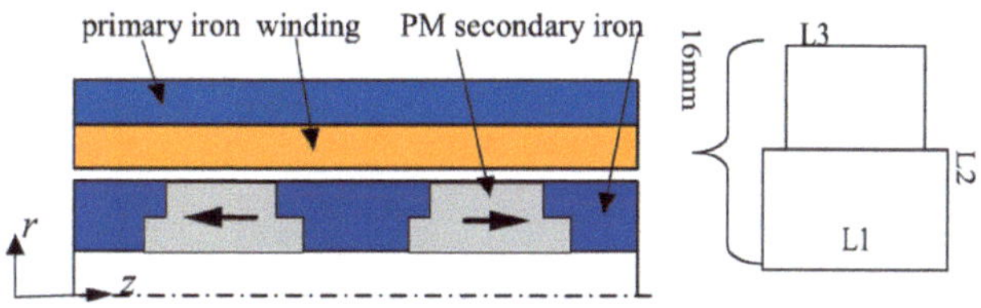

Fig. 3.20 Motor structure in $r - z$ coordinate system in [44]

However, these pole pattern usually cannot improve output performance greatly. Therefore, some researchers focus on linear vernier or magnetic-geared machines [45–49]. These types of linear tubular machines usually incorporate PMs mounted on the stator surface and utilize magnetic gearing principle. In other words, a magnetic gear is integrated in linear tubular machine, and hence it can significantly increase force output while low translation speed. Alternatively, transverse-flux tubular linear machine is another solution that can achieve relatively high force output [50]. Because it has spatially independent magnetic and electrical load and hence achieve large flux and low magnetic reluctance [51]. As shown in Fig. 3.21, Pan et al. [52] proposed a gear-shaped teeth transverse-flux linear machine to shorten magnetic circulation paths for large force output. To overcome the inherent large flux leakage in transverse-flux linear machine, as shown in Figs. 3.22 and 3.23, a double N poles and double S poles pattern and a configuration of consequent-pole mover and segment-interlacing stator are presented in [53, 54], respectively. However, these pole patterns and configurations generally result in complex structure and large volume because of their three-dimensional magnetic circuits or special working principle such as magnetic gearing principle.

To decrease mover mass, the quasi-Halbach array has attracted much attention. It mainly attribute to the fact that quasi-Halbach array has self-shielding effect that can weaken magnetic filed within the back-iron side. Hence, the back-iron thickness can

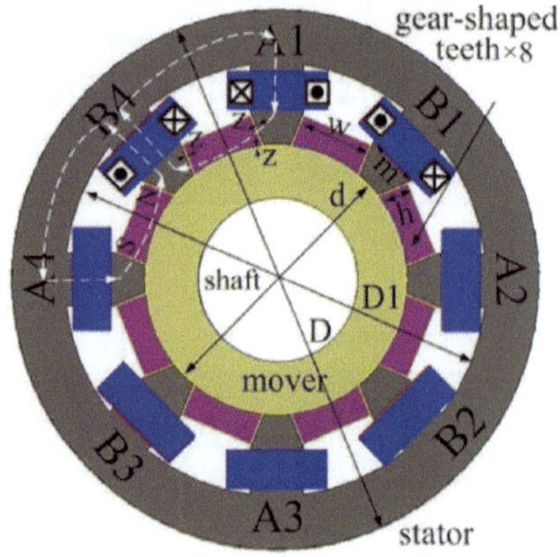

Fig. 3.21 Gear-shaped teeth transverse-flux linear machine in [52]

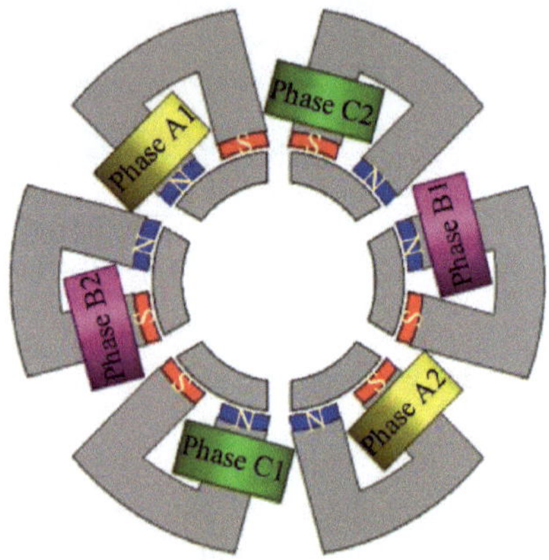

Fig. 3.22 Structure diagram of the TFFRPMLM in [53]

be reduced to a certain extent contributing to the decrease of mover mass. In [55], a yoke-less linear PM motor with quasi-Halbach array are described, as shown in Fig. 3.24, and the utilization of space and machine materialis are improved. As shown in Fig. 3.25, Wang et al. [56] develop a tubular linear machine with quasi-Halbach array for active vehicle suspension. The very thin ferromagnetic tube is employed to enhance the dynamic capability, however, resulting in the reduction of the thrust force due to the incomplete self-shielding effect. The variant of quasi-Halbach, such as T-type, trapezoidal and triangular quasi-Halbach in Fig. 3.26, are investigated for enhancing self-shielding effect, consequently, the mover mass are reduced somehow [57]. In fact, the ideal Halbach can achieve great self-shielding and flux-focusing effect, however, it is very difficult to implement due to the very complicated magnetization direction. Thus, segment-Halbach array becomes a promising candidate as a compromise of ideal Halbach array. But, it also bring a challenge for the fabrication and magnetization technique.

The novel compound Halbach array to simultaneously achieve large force output and low mover mass for tubular linear oscillating machine. The characteristics of flux

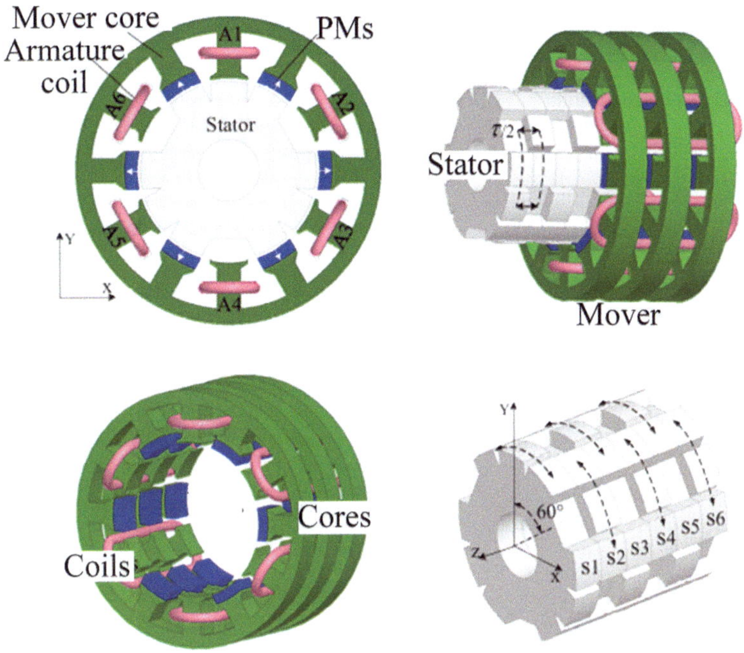

Fig. 3.23 Single-phase configuration of the proposed TF-PMLM in [54]

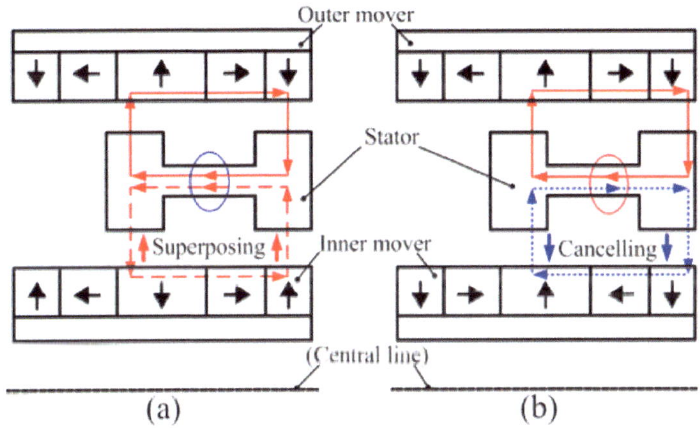

Fig. 3.24 Flux path with different types of Halbach magnetization in [55]

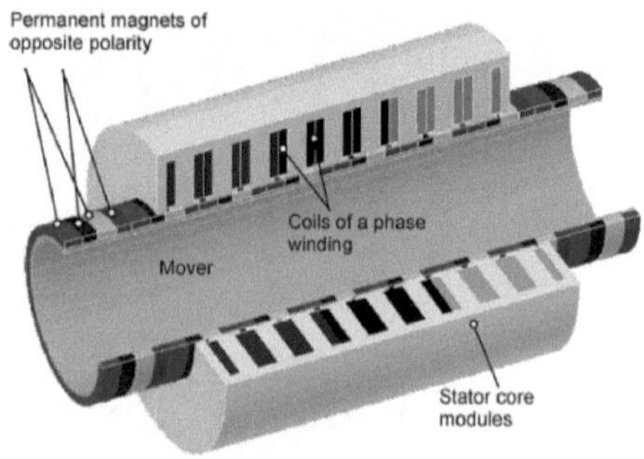

Fig. 3.25 Schematic of nine-slot ten-active-pole tubular PM machine with quasi-Halbach magnetization in [56]

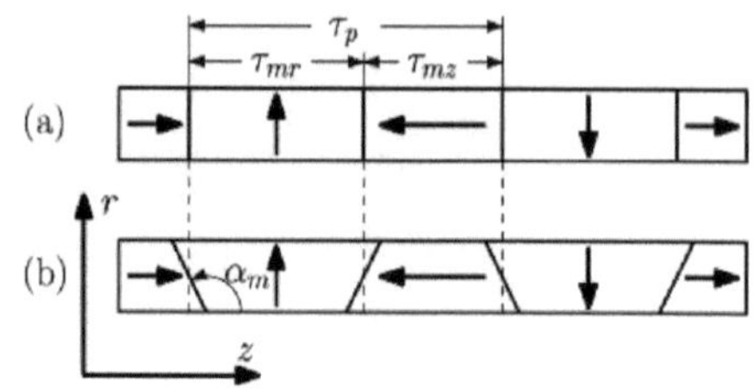

Fig. 3.26 Quasi-Halbach magnetization pattern in [57]

field and force output of the compound Halbach array was present in [36], however, the parameter coupling effects are not considered well. Thus, the response surfaces of force output, thrust fluctuation and energy loss are established for investigating the coupling effect of these key design parameters on output characteristics in this section. Then, the multi-objective genetic optimizations are carried out to obtain the structure parameters. Finally, the electromagnetic performance of the compound Halbach array are compared with conventional Halbach array.

The research content of this chapter is shown in Fig. 3.27. Specifically, Sects. 3.2.1–3.2.5 are introductions to CPMIP machine. Section 3.2.1 presents the schematic structure and pattern analysis of the proposed CPMIP machine. Section 3.2.2 focuses on the mathematical modeling of the magnetic field of the electrical machine. In Sect. 3.2.3, the torque characteristics are investigated and the design improvement is further presented. Then, an electrical machine with irregular PMs and iron poles is designed. In Sect. 3.2.4, the output performance of the electric machine with CPMIP pole patterns is compared with that of the conventional designs. In Sect. 3.2.5, the research prototype is developed, and experiments are conducted to validate the system output performance. Sections 3.3.1–3.3.5 are introductions to novel compound Halbach array. Section 2 presents the schematic structure and working priciple of novel compound Halbach array. Section 3.3.2 presents the analysis of mathematical models. In Sect. 3.3.3, the characteristics of output performance is ivestigated. Section 3.3.4 focus on the sizing optimization. Section 3.3.5 compares the electromagnetic characteristics. And the study is concluded in Sect. 3.4.

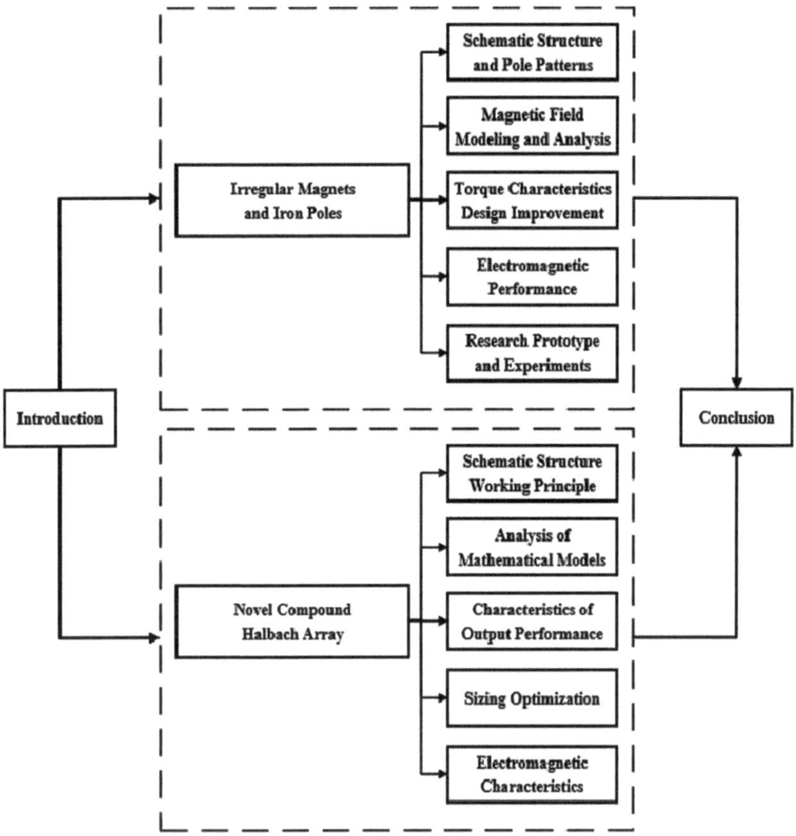

Fig. 3.27 The research content of this chapter

3.2 Irregular Magnets and Iron Pole Pattern Topology

3.2.1 Schematic Structure and Pole Patterns

3.2.1.1 Overall Schematic Structure

Figure 3.28 shows the schematic structure of the proposed CPMIP machine. The configuration of 18-slot/16-pole fractional slot concentrated winding is employed for the machine design to achieve short end winding and small copper volume. Compared with conventional spoke type and other IPM electrical machines, the proposed machine employs irregular magnets and iron poles to achieve strong self-shielding effect, and thus the aluminum can be utilized for the rotor yoke to decrease its mass greatly and improve the system dynamics well. Because there is much less flux loop in aluminum yoke, the eddy current is also depressed greatly. Figure 3.29 presents five pole patterns that could be employed for the rotor design. Figure 3.29a, b and c

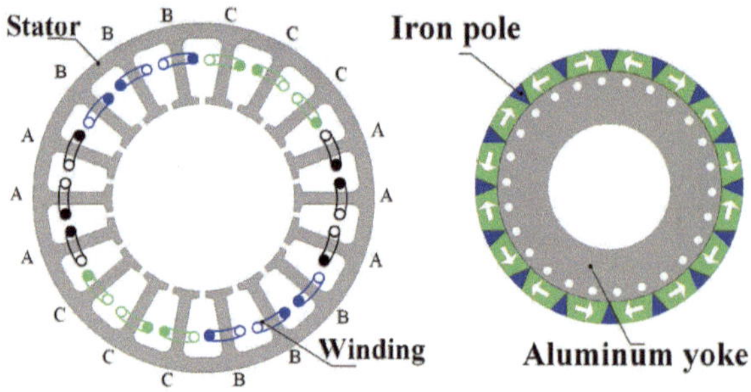

Fig. 3.28 The schematic structure of the proposed CPMIP machine

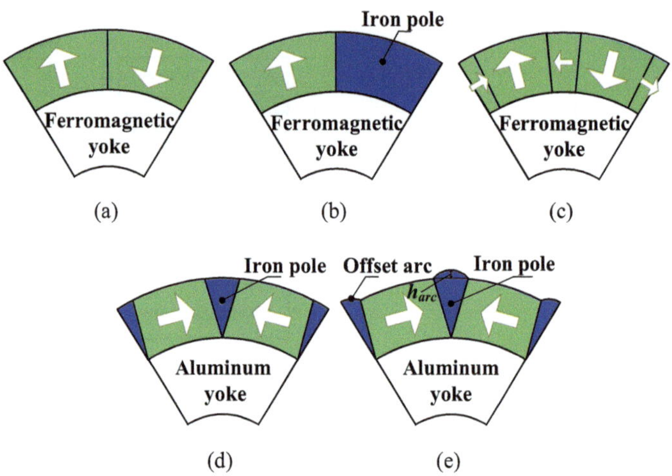

Fig. 3.29 Rotor pole patterns for this study. **a** Radial pole pattern. **b** CPM pole pattern. **c** Halbach array. **d** CPMIP-1 pattern. **e** CPMIP-2 pattern

are the radial magnetization pattern, conventional CPM pattern and Halbach array, respectively. Two types of pole patterns with quasi-triangular iron blocks are proposed in this study. In Fig. 3.29d, the quasi-triangular iron pole has a top arc that is concentric with the rotor surface and has the same radius of the rotor (CPMIP-1). In Fig. 3.29e, the quasi-triangular iron pole has a top arc that has a offset with respect to the rotor center (CPMIP-2). The arc shape can be described by the coefficient k_{arc} defined as the ratio of h_{arc} to airgap length g_{air}. The two machines have same quasi-trapezoid magnets between iron blocks, and the quasi-trapezoid magnet has large cross-section to achieve high airgap flux density, compared with rectangular magnet.

3.2.1.2 Feature Analysis of the Proposed Pole Patterns

The two pole patterns proposed in this study have several advantages as follows.

First, as indicated in Fig. 3.29b, some magnet poles in the conventional CPM pole pattern are replaced with salient iron poles, i.e., the amount of PMs is reduced by half, compared with conventional surface-mounted pole design as shown in Fig. 3.29a. Thus, the magnetomotive force and the torque output may be decreased much. However, in Fig. 3.29d and e, the flux density of each iron pole is generated by two adjacent PMs, thus there is flux-focusing effect that can improve the torque output. The employment of quasi-triangular iron can further enhance this effect and increase the airgap flux density.

Second, the quasi-triangular iron poles can also help to enhance the self-shielding effect. The progressively increased reluctance of iron pole in radial direction leads the flux to the top of iron pole, instead of the rotor center. Thus, the conventional ferromagnetic materials in the rotor can be replaced with low-density materials such as aluminum or polymers. In this case, the rotor mass can be reduced significantly, and the system dynamics can thus be improved. The utilization of aluminum or polymers also benefits the measurement of magnetic flux leakage, as a series of holes are drilled in the aluminum for the placement of gauss probe and would not influence the flux distribution. If polymers are used for the rotor yoke, the core loss can be removed completely in theory.

Third, due to the utilization of iron poles, the PM volume in the rotor is decreased much, which along with the increase of torque output helps to improve the PM utilization ratio, i.e., the ratio of torque output and PM volume [28]. Meanwhile, the reduction of magnet volume can decrease the manufacturing cost of the CPMIP machine as well.

Furthermore, for the conventional pole configuration in Fig. 3.29b, the adjacent PM and iron poles constitute one pole pair, resulting in asymmetric airgap flux distribution and unbalanced magnetic force that in turn may cause serious vibration and noise. In Fig. 3.29d and e, however, the tangentially magnetized quasi-trapezoid poles and quasi-triangular iron poles constitute a sandwich structure, which lead to symmetric airgap flux distribution and balanced magnetic force.

Finally, the proposed configuration provides an effective tool to depress the torque ripple. The flux distribution in airgap can be regulated by the opening size of the quasi-triangle iron pole. The optimization of iron pole's opening angle may achieve high fundamental amplitude and low high-order harmonics of the flux field in airgap, and thus the torque ripple can be decreased. The above-mentioned features will be validated in subsequent sections.

3.2.2 *Magnetic Field Modeling and Analysis*

3.2.2.1 Mathematical Modeling of Flux Field

Figure 3.30 illustrates the equivalent magnetic circuits of three pole patterns, i.e., the conventional CPM pattern, Halbach array and the proposed CPMIP-2 pole pattern. It is worth noting that the magnetic circuits of radial pole pattern and CPMIP-1 is similar to that of CPM and CPMIP-2 patterns, respectively. As shown in Fig. 3.30a, the CPM pole pattern has regular flux distribution and PM shape. Thus the rotor yoke, surface-mounted PMs, iron poles and airgap can be approximately divided as rectangle meshes and their permeances are

$$\begin{cases} G_{radial\_yoke} = 2\mu_0 L_a \tau_p (R_m - 0.5h_m)/(\pi R_r \tau_p) \\ G_{radial\_pm} = \mu_{pm} L_a \tau_p (R_m - 0.5h_m)/h_m \\ G_{radial\_iron} = \mu_{iron} L_a \tau_p (R_m - 0.5h_m)/h_m \\ G_{radial\_ag} = \mu_0 L_a \tau_p (R_m + 0.5g_{air})/g_{air} \end{cases} \quad (3.1)$$

Fig. 3.30 Magnetic circuit models. **a** Radial or CPM pattern. **b** Halbach array. **c** CPMIP-1 or CPMIP-2 pattern

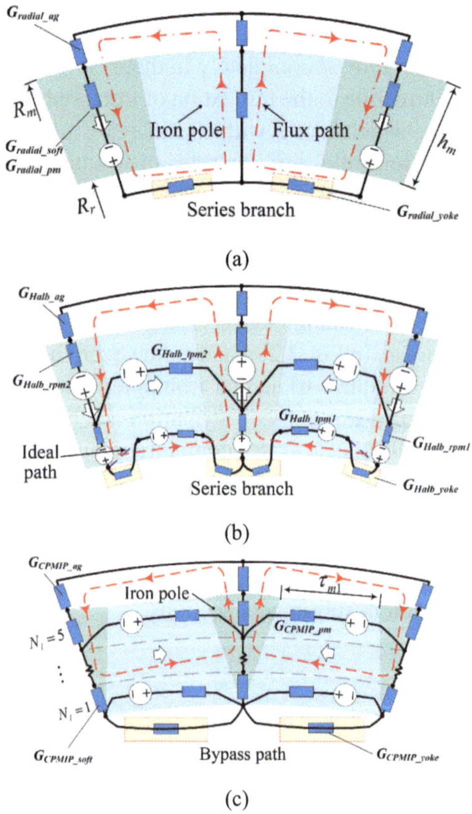

where $\mu_{pm} = \mu_0\mu_{rpm}$ and $\mu_{iron} = \mu_0\mu_{ironpole}$; μ_0, $\mu_{ironpole}$ and μ_{rpm} are the vacuum permeability and the relative permeability of iron pole and PMs, respectively; τ_p is the pole pitch; R_m and R_r are the inner and outer radius of PMs; h_m and g_{air} are the PM thickness and airgap width, respectively. The magnetomotive force of PMs can be obtained from $F_{radial\_pm} = H_c h_m$, where H_c is the coercivity of magnets.

As shown in Fig. 3.30b, the Halbach PMs are divided into two layers that can be approximated as rectangles. The permeances of these parts are

$$\begin{cases} G_{Halb\_yoke} = 0.5\mu_0 L_a \tau_p R_r/(0.25\tau_p R_r \pi) \\ G_{Halb\_tpm1} = 0.5\mu_{pm} L_a h_m/(\tau_p(R_r + 0.125h_m)) \\ G_{Halb\_tpm2} = 1.5\mu_{pm} L_a h_m/(\tau_p(R_m - 0.375h_m)) \\ G_{Halb\_rpm1} = \tau_p \mu_{pm} L_a (R_r + 0.125h_m)/(0.5h_m) \\ G_{Halb\_rpm2} = \tau_p \mu_{pm} L_a (R_m - 0.375h_m)/(1.5h_m) \\ G_{Halb\_ag} = \mu_0 L_a \tau_p (R_m + 0.5g_{air})/g_{air}. \end{cases} \tag{3.2}$$

where G_{Halb_yoke}, G_{Halb_ag}, G_{Halb_tpm} and G_{Halb_rpm} are the permeances of rotor yoke, airgap and PMs of tangential and radial magnetization, respectively. The magnetomotive force of Halbach array can be obtained from the following equations, $F_{H\_rm1} = 0.25H_c h_m$, $F_{H\_tm1} = 0.5H_c\tau_p(R_r + 0.125h_m)$, $F_{H\_rm2} = 0.75H_c h_m$, $F_{H\_tm1} = 0.5H_c\tau_p(R_m - 0.375h_m)$.

As shown in Fig. 3.30c, the equivalent magnetic network is established for the proposed CPMIP pole pattern. The quasi-trapezoid PMs and quasi-triangular iron poles are divided into five layers to improve computational precision and efficiency. The width and height of the approximative rectangles of different layers vary with the number of layers N_l. The corresponding permeances of subparts are expressed with

$$\begin{cases} G_{CPMIP\_yoke} = \mu_0 L_a h_m/(N_l \tau_p (R_r - 0.125hm)) \\ G_{CPMIP\_soft} = N_l \mu_0 \mu_{rsoft} L_a (\tau_p - \theta(i))R(i)/h_m \\ G_{CPMIP\_pm} = \mu_{pm} L_a h_m/(N_l R(i)\theta(i)) \\ G_{CPMIP\_ag} = \mu_0 L_a \tau_p (R_m + 0.5g_{air})/g_{air} \end{cases} \tag{3.3}$$

where $i \in [1, 5], \theta(i) = \tau_{m2} - 0.1(\tau_{m2} - \tau_{m1})(2i - 1), R(i) = R_r + 0.1h_m(2i - 1)$, $\tau_{m2} = \tau_p$ and $\tau_{m1} = k_{open}\tau_p$; μ_{rsoft} is the relative permeability of the quasi-triangular iron pole. The coefficient k_{open} determines the opening angle of triangular iron pole. In addition, the magnetomotive force of magnets can be obtained from $F_{CPMIP\_pm} = H_c R(i)\theta(i)$. The magnetic circuit model can be expressed with matrix equation as

$$\boldsymbol{G} \cdot \boldsymbol{F_{node}} = \boldsymbol{\Phi} \tag{3.4}$$

where $\boldsymbol{G}$ is the permeance matrix directly determined by (3.1), (3.2) and (2.1); $\boldsymbol{F_{node}}$ and $\boldsymbol{\Phi}$ are the magnetomotive force matrix and flux matrix, respectively. The representation of these matrices is closely related to the node arrangement shown in Fig. 3.30. These matrices can be expressed as

$$\boldsymbol{G} = \begin{bmatrix} G(1,1) & \cdots & G(1,N_{node}) \\ \vdots & \ddots & \vdots \\ G(N_{node},1) & \cdots & G(1,N_{node}) \end{bmatrix} \tag{3.5}$$

$$\boldsymbol{F_{node}} = \left[F\,(1) \cdots F\,(N_{node}) \right] \tag{3.6}$$

$$\boldsymbol{\Phi} = \left[\Phi\,(1) \cdots \Phi\,(N_{node}) \right] \tag{3.7}$$

where N_{node} is the node number of magnetic circuit model. $\boldsymbol{G}$ is a symmetric positive definite matrix with the relation of

$$\begin{cases} G(i,i) \geq 0, G(i,j) \leq 0 & i \neq j \\ G(i,i) = -\sum_{j=1,j\neq i}^{N_{node}} G(i,j) & i = j \\ G(i,j) = G(j,i) & i \neq j \end{cases} \tag{3.8}$$

By solving the nonhomogeneous linear equations in (2.2) of three magnetic circuit models, the magnetomotive force of each node, $F(i)$, can be obtained. Then the flux density is

$$B = (F(i) - F(j))G(i,j)/S_{ele} \tag{3.9}$$

where S_{ele} is the equivalent cross-section area of $G(i,j)$ perpendicular to the flux direction.

3.2.2.2 Magnetic Flux Field Analysis

Figure 3.31a and b show the analytical and numerical results of flux density in air-gap and rotor yoke, respectively. The root mean square values for three types of arrays, i.e., CPM pole pattern, Halbach array and CPMIP pole pattern, are compared. Both ferromagnetic and aluminum yokes are considered. The figures indicate satisfactory agreement between analytical and numerical computation, which verifies the magnetic circuit models. In addition, the magnetic circuit models shown in Fig. 3.30 illustrate that the high-density yoke is necessary for radial, CPM and Halbach patterns to reduce magnetic resistance through the ferromagnetic yoke (series branch).

From the figures, the airgap flux density of the CPMIP pole pattern is larger than that of CPM with high-density ferromagnetic yoke, by 50.17%. In other words, the flux-focusing effect of CPMIP pole pattern is significant due to the interaction of neighboring PMs, high magnetomotive force of quasi-trapezoid magnets and the large cross section area of the quasi-triangular iron pole, which contribute to the flux increase of the magnetic circuit.

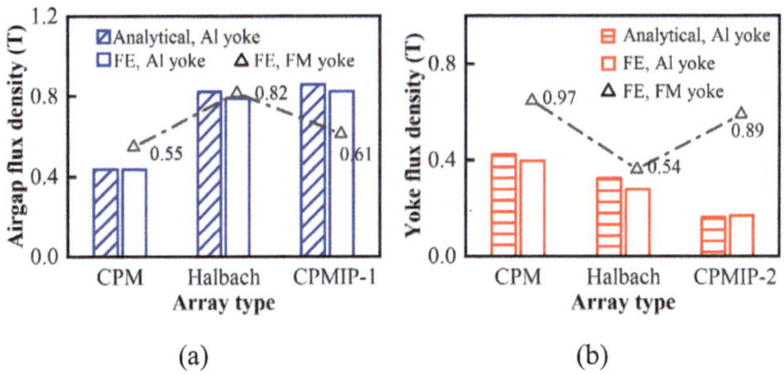

Fig. 3.31 Comparison of analytical and numerical results with aluminum or FM yoke. **a** Airgap flux density. **b** Rotor yoke flux density

The influence of iron yoke on the magnetic field of CPM is significant. When the iron yoke is removed, the flux density in the airgap is reduced by 20.17%. As shown in Fig. 3.30, there are series of magnetic flux branches including the radially magnetized PMs in the rotor yoke. Removing the ferromagnetic yoke increases the branch reluctance, and hence decreases the airgap flux density for the conventional CPM pole pattern and Halbach array. By contrast, this branch does not exist in the CPMIP patterns, and the flux density of the CPMIP is even larger when there is no iron yoke. Figure 3.31b indicates that the flux density in the yoke of CPMIP is much smaller than that of CPM and Halbach patterns, i.e., the self-shielding effect of CPMIP is very impressive. Therefore, the rotor yoke can be removed completely, or replaced with low-density materials. The rotor yoke mass can be reduced by 63.53% if aluminum is utilized, and even more if polymers are employed. The reduction of rotor mass can contribute to the enhancement of torque density and dynamic characteristics.

3.2.3 Torque Characteristics and Design Improvement

3.2.3.1 Torque Characteristics

It can be deduced from (3.3), (3.4) and (3.9) that the airgap flux field distribution is directly related to k_{open}. The armature field varies with the current angle β, and causes the waveform distortion of airgap flux field. It results in different flux harmonics and reluctance torque that may cause the variations of torque ripple, average torque and PM utilization ratio.

Figures 3.32 and 3.33 show the response surfaces of average torque and PM utilization ratio for CPMIP-1 pattern, respectively. It is found that the maximum torque and PM utilization ratio appear at $k_{open} = 0.7$ and $k_{open} = 0.2$, respectively, with $\beta = 8$. Large k_{open} indicates thick PM and small opening size of iron pole which in turn results in strong magnetomotive force and flux-focusing effect. But when k_{open}

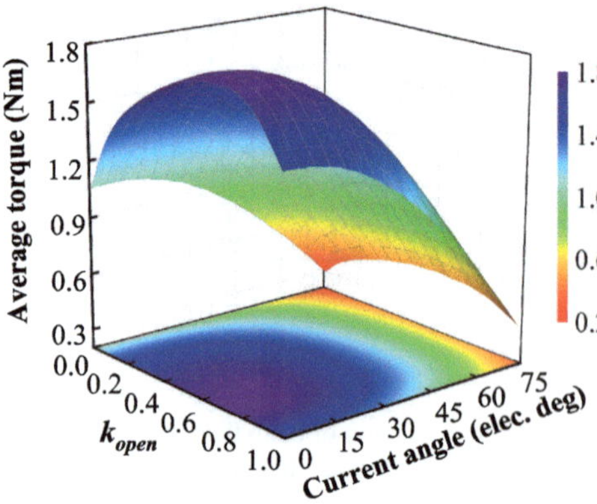

Fig. 3.32 Variation of average torque w.r.t k_{open} and β (CPMIP-1)

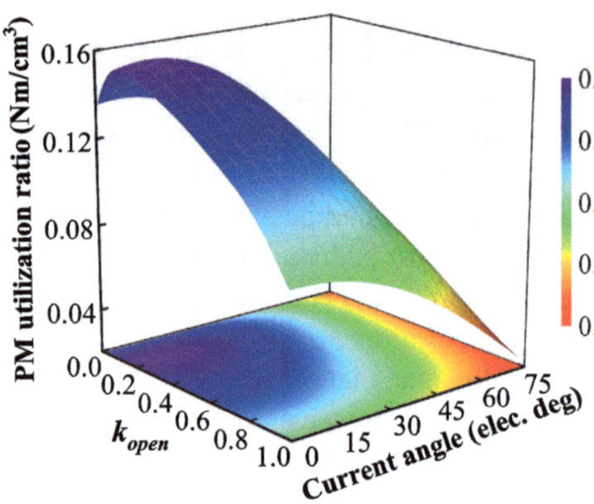

Fig. 3.33 Variation of PM utilization ratio w.r.t k_{open} and β (CPMIP-1)

is too large, the interaction between the two oppositely and tangentially magnetized trapezoidal PMs changes the working point of BH curve of PMs, similar to field weakening effect. In this case, the average torque output is reduced.

Figure 3.34 shows the variation of torque ripple with respect to k_{open} and β in CPMIP-1 pole pattern. It can be seen that the minimum torque ripple appears at $k_{open} = 0.5$ and $\beta = 0$. When k_{open} is small, the direction of the third-order harmonic is opposite to that of the fundamental waveform, which results in a quasi-square waveform. When k is large, its direction is consistent with the fundamental waveform, which results in sharp airgap flux wave. In addition, large current angle β may cause more high-order harmonics. These factors generally lead to large torque ripple.

To solve above problems and improve the output performance of CPMIP-1 further, an alternative design option with offset iron-pole arc, i.e., CPMIP-2 pole pattern shown in Fig. 3.29e, is proposed. Figure 3.35 presents the variation of average torque with respect to k_{open} and k_{arc} for CPMIP-2 pole pattern. It is found that the maximum torque appears in the area where k_{open} is small, i.e. PM volume is small. In other words, the improved design can effectively improve torque output and PM utilization ratio simultaneously. Meanwhile, the torque ripple can be reduced too.

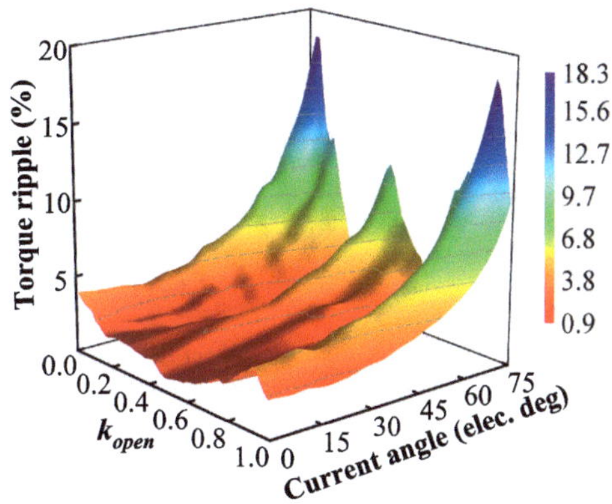

Fig. 3.34 Variation of torque ripple w.r.t k_{open} and β (CPMIP-1)

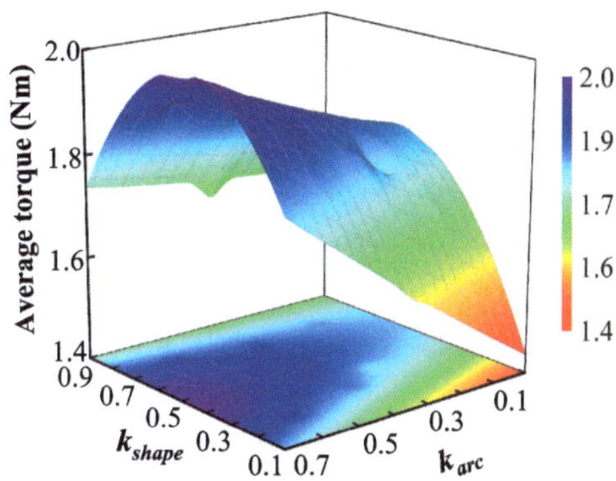

Fig. 3.35 Variation of average torque w.r.t k_{open} and k_{arc} (CPMIP-2)

3.2.3.2 Design Process

The design constraints are listed in Table 3.1. The major design variables are the inner radius of the stator, the airgap length, the number of slots and pole pairs,

Table 3.1 Design constraints and the final values

Symbol	Quantity	Value	Final value
B	Magnetic flux density(max)	<1.8 T	1.7 T
J	Current density	<6 A/mm^2	5 A/mm^2
g	Airgap length	<1mm	0.8 mm
D_o	Outer diameter	<90 mm	84 mm
l_a	Active stack length	$\leq$ 40 mm	40 mm
T_{avg}	Average torque	>1.8 Nm	1.907 Nm
V_{pm}	Total magnet volume	<12 cm^3	11.15 cm^3
M_{rotor}	Rotor mass	<0.3 kg	0.249 kg

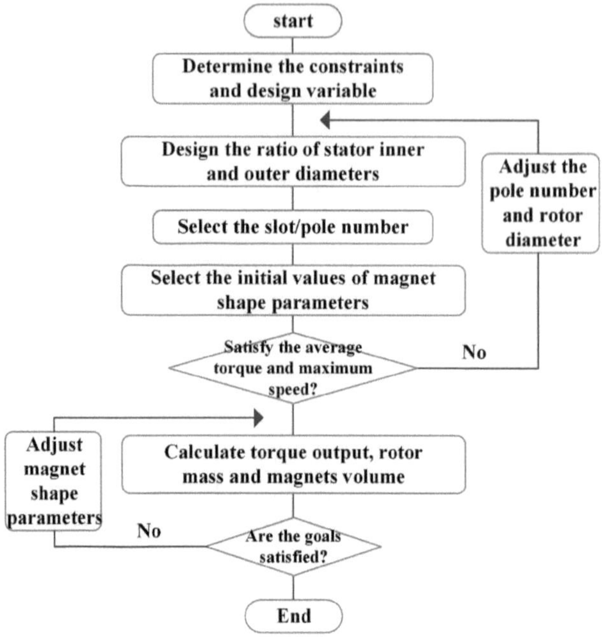

Fig. 3.36 Design process flowchart for CPMIP machine

and the structure parameters of the quasi-cylindrical magnets, i.e., k_{open}, k_{arc} and the magnet length. The design process is presented in Fig. 3.36. Specifically, when the outer diameter of stator core has been determined, the stator inner diameter which significantly affect torque output should be considered. A large stator inner diameter helps to improve average torque because of the large moment arm and high airgap flux density. However, it may result in magnetic saturation in stator tooth and high current density. Following that, the slot/pole number should be determined. To obtain low torque ripple and short end winding, a large slot/pole number of fractional slot concentrated winding is considered, while the value of the slot/pole number is limited by the required maximum rotor speed. Thus, according to the required average torque, the maximum speed and the flux density in stator teeth, the values of the slot/pole number and stator inner diameter are selected. To reduce total magnet volume while maintaining high average torque, the quasi-triangular magnet structure parameters, such as k_{open}, k_{arc} and magnet length are determined through design iteration and numerical validation. At the same time, to ensure that the removal of high-density ferromagnetic rotor yoke for low rotor mass would not influence average torque, the self-shielding effect related to the magnet structure parameters should be considered in the iteration. The final values are listed in Table 3.1 and the design specification is listed in Table 3.2.

Table 3.2 Key design specifications of the investigated pole patterns

Items	Radial	Halbach	CPM	CPMIP-1	CPMIP-2
Number of phases	3	3	3	3	3
Outer diameter of stator (mm)	84	84	84	84	84
Inner diameter of stator (mm)	45	45	45	45	45
Stator tooth width (mm)	3	3	3	3	3
Slot opening length (mm)	1.5	1.5	1.5	1.5	1.5
Air-gap length (mm)	0.8	0.8	0.8	0.8	0.8
Active stack length, l_a (mm)	40	40	40	40	40
PM length (mm)	3.5	3.5	3.5	3.5	3.5
Opening coefficient, k_{open}	–	–	–	0.63	0.34
Iron-pole arc coefficient, k_{arc}	–	–	–	–	0.6
Current angle β (deg)	0	0	10	8.405	15.76
PM grade	N42UH	N42UH	N42UH	N42UH	N42UH
Turns per coil	15	15	15	15	15
Rated current (A)	6	6	6	6	6
Rated DC bus voltage (V)	48	48	48	48	48
Rated speed (r/min)	1500	1500	1500	1500	1500

3.2.4 Comparison of Electromagnetic Performance

In this section, the electromagnetic performance of the proposed CPMIP-1 and CPMIP-2 pole patterns is compared with that of the conventional CPM, Halbach and radial magnetization patterns, including open-circuit flux density, torque output, PM utilization, rotor mass, unbalanced magnetic force and efficiency. For a fair comparison, these pole patterns have the same stator and rotor dimension, magnet length, rated speed and input current, as shown in Table 4.5.

3.2.4.1 Open-Circuit Flux Density in Airgap and Rotor Yoke

Figure 3.37 shows the magnetic field of the five designs. It is clearly found that the machines with CPMIP pole patterns shown in Fig. 3.37e and f have strong flux-focusing effect, i.e., high flux density in airgap and stator core, and strong self-shielding effect, i.e., very low flux density in rotor yoke. The spoke-type interior PM machine is also investigated and shown in Fig. 3.37d for reference. It is found that the tangentially magnetized rectangular permanent magnets in the spoke type pole pattern have unipolar flux leakage even if the air yoke is used, which may result in lower average torque. Figure 3.38 shows the open-circuit flux density waveform

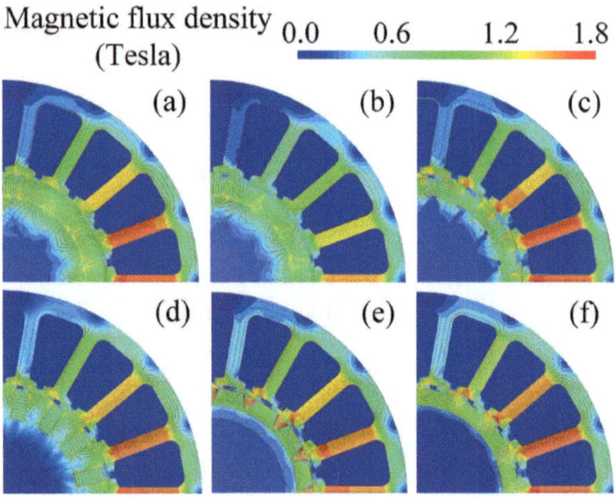

Fig. 3.37 Magnetic field distribution. **a** Radial pole pattern. **b** CPM rotor pole pattern. **c** Halbach array. **d** Spoke-type design. **e** Proposed CPMIP-1 pole pattern. **f** Proposed CPMIP-2 pole pattern

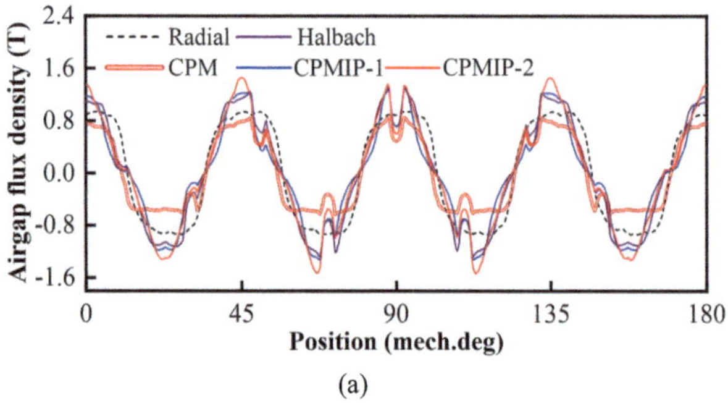

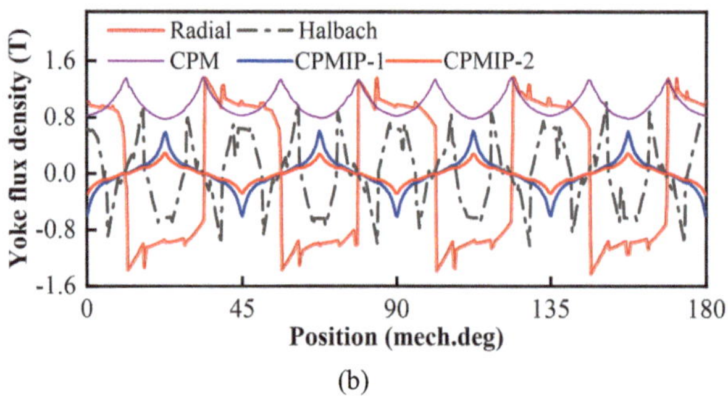

Fig. 3.38 Flux density distribution. **a** Airgap. **b** Rotor yoke

in airgap and rotor yoke, sequentially. From Fig. 3.38a, it can be clearly found that the CPM pattern has asymmetric airgap flux density wave (even-order harmonics) and lowest fundamental term. On the contrary, the proposed CPMIP patterns have symmetric flux wave, and the fundamental term of airgap flux density is higher, compared with the conventional CPM and radial patterns. In addition, the CPMIP-2 pole pattern has highest airgap flux density amplitude among these five pole patterns due to the strong flux-focusing effect. Besides, the CPMIP patterns also have smallest flux density within the rotor yoke shown in Fig. 3.38b, which indicates that the self-shielding effect of CPMIP pole patterns are better than that of other three patterns. Thus the CPMIP machine rotor yoke with high-density ferromagnetic material can be completely removed to reduce the rotor mass.

3.2.4.2 Comparison of Torque Performance

The average torque output of these pole patterns are investigated and compared. The investigated machines with different pole patterns are compared under the optimal current angles. The results are presented in Fig. 3.39. The average torque output of the conventional CPM pattern is reduced by 20.82% compared with the radial pattern, although it uses less PM volume. On the contrary, the proposed CPMIP-1 and CPMIP-2 have 2.22% and 8.48% higher average torque, respectively, than the radial pattern due to their strong flux-focusing effect. The average torque output of CPMIP-2 pattern is higher than that of Halbach array. In addition, CPMIP-1 and CPMIP-2 has 60% and 62.86% lower torque ripple than the radial pattern, and the CPMIP-2 has the smallest torque ripple among these pole patterns, as listed in Table 3.3. The above comparisons demonstrate that the quasi-triangular iron poles can improve the flux distribution in airgap and the torque output well. The reluctance torque by the iron poles can also be utilized to increase the torque further. In other words, the iron pole brings a comparable saliency ratio, i.e., the iron pole can reduce the q-axis magnetic reluctance, which results in the difference between the inductances of d- and q-axis.

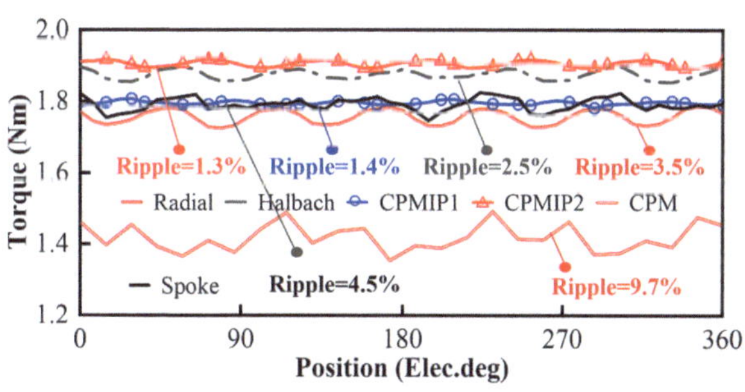

Fig. 3.39 Variation of torque output with respect to rotor position

Table 3.3 Torque characteristics and magnet utilization

Items	Radial	Halbach	CPM	CPMIP-1	CPMIP-2
T_{ave}(Nm)	1.758	1.876	1.417	1.797	1.907
$T_{ave}[ratio]$	100%	106.7%	80.6%	102.2%	108.4%
T_{ripple}(%)	3.5	1.4	9.7	1.4	1.3
$T_{ripple}[ratio]$	100%	40%	277%	40%	37.14%
V_m(cm^3)	17.549	17.549	8.775	14	11.153
η_m(Nm/cm^3)	0.1	0.107	0.159	0.128	0.171
$\eta_m[ratio]$	100%	106.7%	158.4%	128.1%	170.7%
M_{rotor}(kg)	0.450	0.450	0.451	0.221	0.258
$M_{rotor}[ratio]$	100%	100%	100.22%	49.11%	57.33%

For a more comprehensive comparison, the torque performance of the spoke-type pole pattern with the same magnet volume and stator structure is studied. The average torque and the torque ripple of the spoke-type design are 1.79Nm and 4.51%, respectively. The average torque and the torque ripple of the CPMIP-2 pole pattern are 1.907Nm and 1.3%, respectively. In other words, the average torque is improved by 6.5%, and the torque ripple is reduced by 71.2%, because the tangentially magnetized rectangular magnets in the spoke-type design have large unipolar flux leakage even if the air yoke is used as shown in Fig. 3.37. Besides, the rotor mass of the CPMIP-2 pole pattern and the spoke-type pole pattern is 0.258kg and 0.399kg, respectively. In other words, the rotor mass is reduced by 35.34% with the CPMIP-2 pole pattern rotor.

3.2.4.3 Comparison of PM Utilization and Rotor Mass

PM is one major material cost of electrical machines. It can be measured by the PM volume V_m and PM utilization ratio η_m in Table 3.3 defined by the ratio of average torque and PM volume. The CPMIP-1 and CPMIP-2 patterns reduce the PM volume by 20.23% and 36.45% than the conventional pole patterns. In addition, the CPMIP-2 provides the highest PM utilization ratio, specifically, 170.7% of the radial pattern, 159.8% of the Halbach array and 107.6% of the CPM pattern.

It is worth noting that while CPMIP-1 and CPMIP-2 improve the average torque and PM utilization ratio, the rotor mass of CPMIP-1 and CPMIP-2 are only 49.11% and 57.33% of the conventional pole patterns.

3.2.4.4 Unbalanced Magnetic Force

The unbalanced magnetic force (UMF) is undesirable because it is one of major sources of vibration and noise. It has been proved that the UMF does not exist in the conventional fractional-slot PM machine with $N_s = 2p \pm 2$, but it may occur in consequent-pole design even with this special slot-pole configuration due to the asymmetric airgap flux density [28, 29]. Figure 3.40 is one typical example showing the radial force density wave for consequent-pole design. The asymmetrical wave usually causes UMF on the electrical machine.

The force components of UMF, F_x and F_y, acting on a rotor are given by

$$F_x = \frac{rl_a}{2\mu_0} \int_0^{2\pi} [(B_\alpha^2 - B_r^2)\cos\alpha + 2B_r B_\alpha \sin\alpha]d\alpha \tag{3.10}$$

$$F_y = \frac{rl_a}{2\mu_0} \int_0^{2\pi} [(B_\alpha^2 - B_r^2)\sin\alpha - 2B_r B_\alpha \cos\alpha]d\alpha \tag{3.11}$$

where B_α and B_r are airgap flux density in tangential and radial directions, respectively. Figure 3.41 shows the unbalanced magnetic force of the CPM pole pattern, the conventional Halbach array and the CPMIP pole patterns. It is found that the

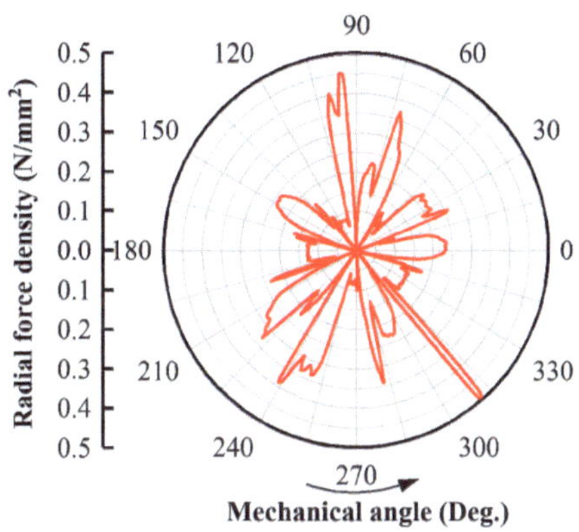

Fig. 3.40 Radial force density wave in CPM pattern machine

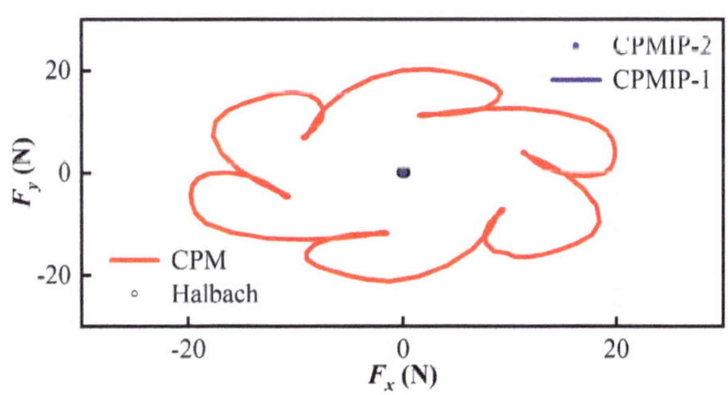

Fig. 3.41 Unbalanced magnetic force of the studied pole patterns

unbalanced magnetic force of the CPMIP patterns and the Halbach array is much lower than that of the CPM pole pattern, and both are close to zero and overlapped in the figure.

3.2.4.5 Loss and Efficiency

The PM eddy loss of the five machine designs is shown in Fig. 3.42. The results indicate that the proposed CPMIP-2 pole pattern has the lowest PM eddy loss. Compared with Halbach array, the PM loss of the CPMIP-2 pole pattern is reduced by 81.16%, because the magnet volume employed in the design is smaller. Besides, the total eddy loss of the CPMIP-1 pole pattern is the lowest. The efficiency maps of the five machines along with the spoke-type design are investigated under the constant power input, and the results are shown in Fig. 3.43. The Halbach array and the proposed CPMIP pole patterns have the highest torque output as shown in Fig. 3.43c, e and f. And the high efficiency area of the CPMIP pole patterns is larger than that of the Halbach array.

3.2.5 Research Prototype and Experiments

3.2.5.1 Development of Research Prototype

The research prototype of electrical machine with CPMIP-2 pole pattern and the parameters of $k_{open} = 0.67$ and $k_{arc} = 0.29$ has been developed. The fractional slot concentrated winding of 18-slot/16-pole is employed for the prototype to reduce the size of end windings. Figure 3.44a shows the output shaft of the developed machine, and it is mounted with low-density aluminum rotor yoke. The rotor and stator are presented in Fig. 3.44b and c, respectively. The quasi-trapezoidal PM can be parallel-magnetized in order to reduce the complexity of magnetization and thus further decrease the manufacturing cost. The assembled prototype is presented in Fig. 3.44d.

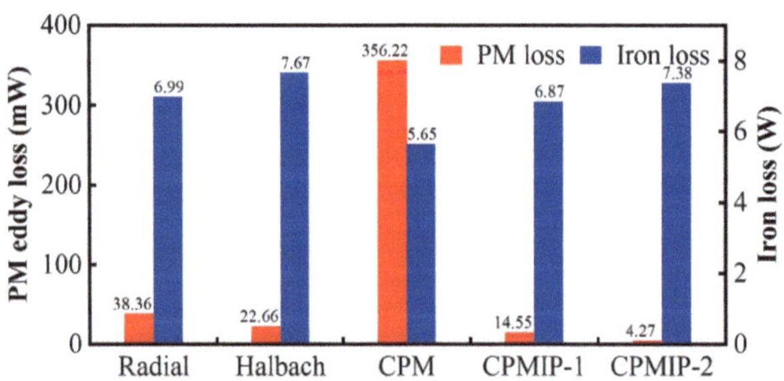

Fig. 3.42 PM eddy loss and iron loss of the five pole patterns

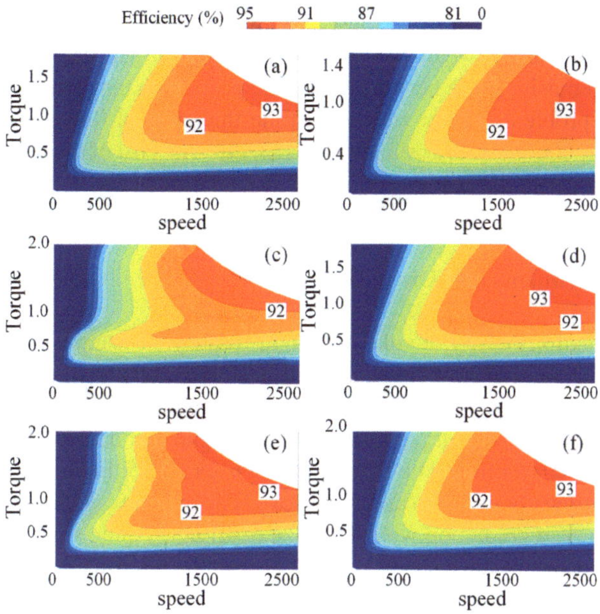

Fig. 3.43 Efficiency maps. **a** Radial pole pattern. **b** CPM pattern **c** Halbach array. **d** Spoke-type design. **e** CPMIP-1. **f** CPMIP-2

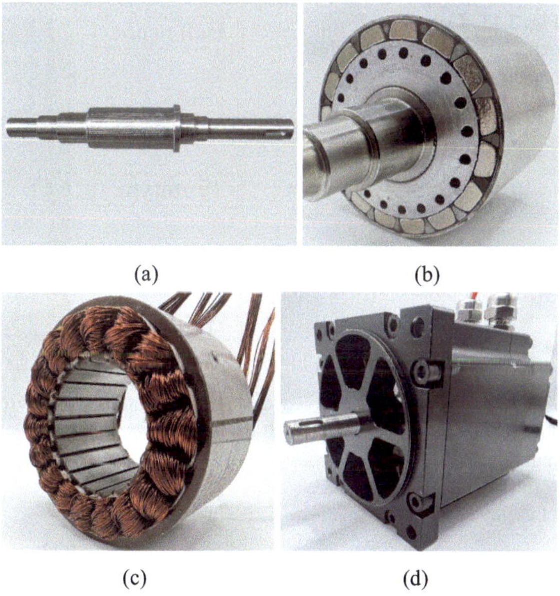

Fig. 3.44 Developed prototype. **a** Output shaft **b** Rotor. **c** Stator core and windings. **d** Assembly

One challenge is the assembly of these irregular magnets and irons. The triangular irons are sandwiched between two adjacent trapezoidal magnets, which causes relatively large repulsive force between magnetic poles with tangential magnetization. To solve this problem, a rotor sleeve with the thickness of 0.4mm is employed to balance this repulsive force, as shown in Fig. 3.44b. The cost of the quasi-trapezoidal magnets is similar to the cost of commonly used tile magnets.

3.2.5.2 Experiments and Discussion

To measure the flux density of yoke region, the rotor yoke is equally spaced with 20 holes with a diameter of 2mm, as shown in Fig. 3.44b. The probe of Gauss meter can be inserted into those holes to measure the flux leakage. Similarly, the probe of Gauss meter can be inserted into airgap for measuring airgap flux density distribution. The constructed test rig for magnetic field is shown in Fig. 3.45. Figure 3.46 presents the measured flux density in airgap and rotor yoke. The experimental result is consistent with the numerical prediction well. The size of airgap is quite small, about 0.3mm, due to the influence of rotor sleeve. Only the radial flux density component near the slot opening can be measured with the 0.25mm Hall probe. It can be found that there is almost no flux leakage in the rotor yoke, which validates its strong self-shielding effect.

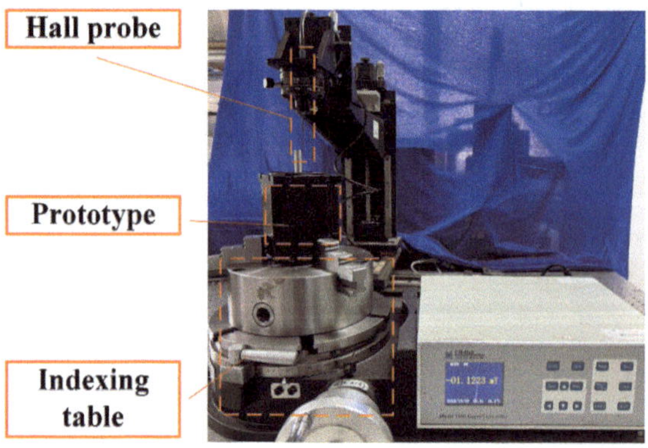

Fig. 3.45 Constructed experimental test-rig for flux density

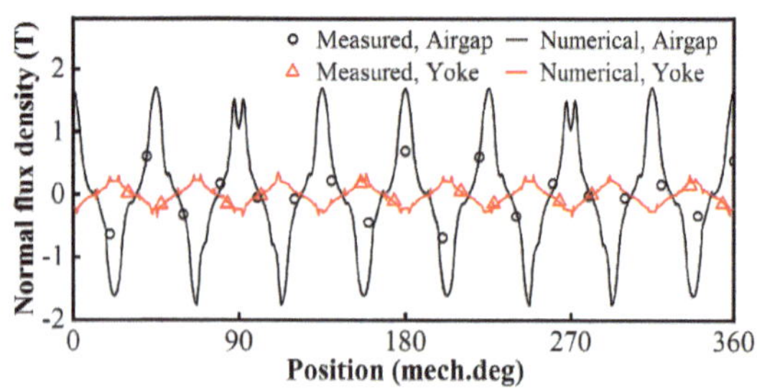

Fig. 3.46 Comparison of numerical and measured results of normal flux density of airgap and yoke

Figure 3.47 shows the comparison of numerical and measured results of line to line back electromotive force (EMF) voltage at the rated speed 1500 r/min. The measured result of back EMF matches the numerical results well.

The test-rig for dynamic torque output is constructed as shown in Fig. 3.48. The research prototype is mounted on a flange which in turn is fixed on the experimental platform. The output shaft of the research prototype is connected to a hysteresis dynamometer through a disc coupling. The prototype generates the rotary motion and pulls the dynamometer to rotate synchronously. The dynamometer can produce loading torque on the prototype. Therefore, the output performance of the electrical machine can be measured accordingly. Figure 3.49 shows the value of torque output

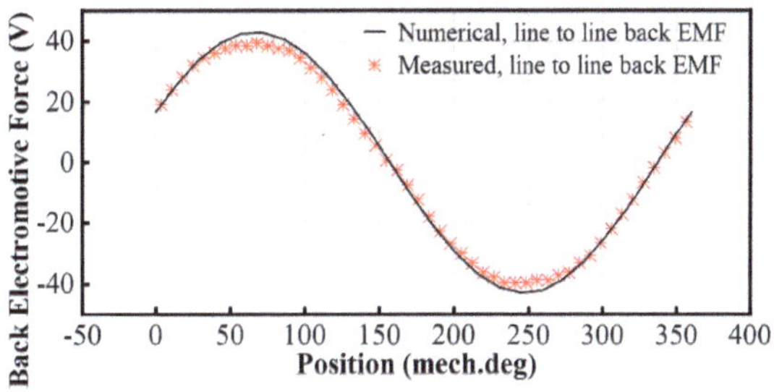

Fig. 3.47 Comparison of numerical and measured results of back EMF

Fig. 3.48 Constructed experimental test-rig for output performance

Fig. 3.49 Comparison of torque output against input current curve

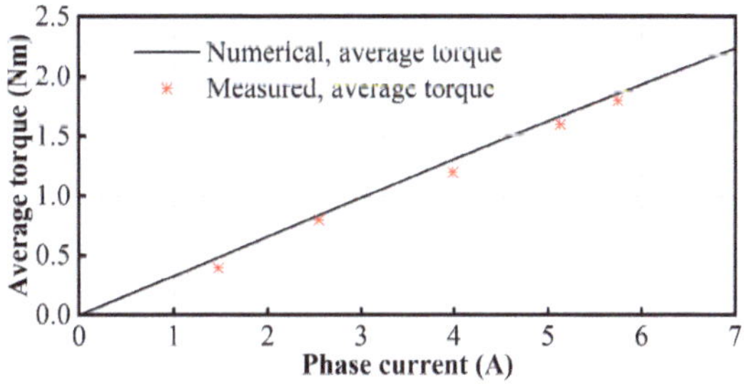

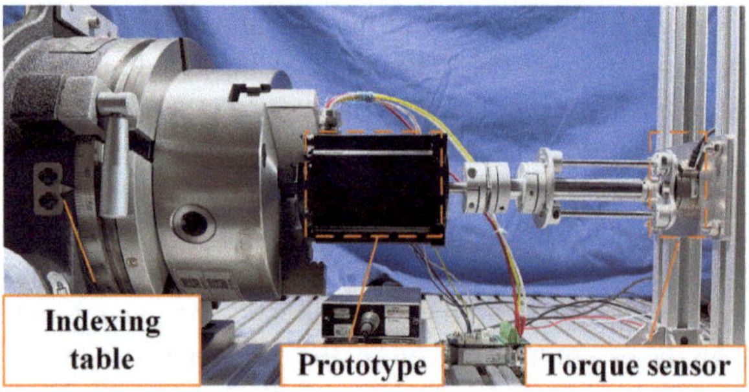

Fig. 3.50 Testing platform for steady torque measurement of the research prototype with the CPMIP pole pattern

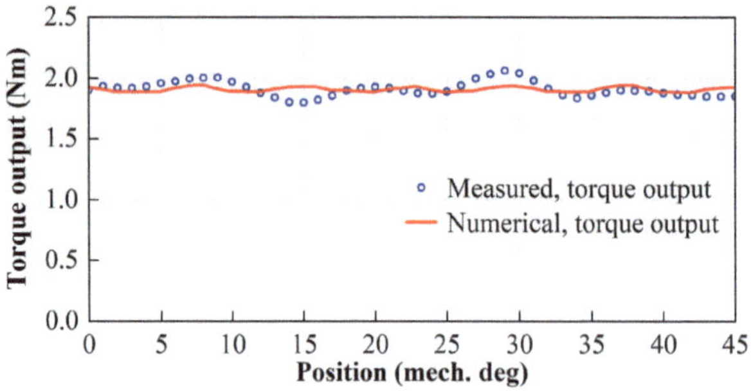

Fig. 3.51 Steady torque output waveform of the research prototype with the CPMIP pole pattern versus its rotor position

against the current input. The experimental measurement result is consistent with the numerical computation.

The measurement platform of the steady torque is developed as shown in Fig. 3.50. One end of the research prototype of the electrical machine is clamped on the indexing table, and the other end is connected with the torque sensor by a mechanical coupling. The indexing table drives the stator to rotate by the step of one degree, and then the controller gives the corresponding excitation current to the three-phase windings according to the position feedback. The steady torque at every position is measured by the torque sensor. Figure 3.51 presents the measured and numerical torque output curves with respect to the rotor position under rated condition. The results validate the proposed machine design and its output performance well.

3.3 Novel Compound Halbach Array Pole Pattern Topology

3.3.1 Schematic Structure and Working Principle

3.3.1.1 Schematic Structure

The schematic structure of the conventional quasi-Halbach array and the compound Halbach array are shown in Fig. 3.52a and b, respectively. There are two layers in the compound Halbach array, in which the out layer and inner layer are conventional quasi-Halbach array and axially magnetized pattern, respectively. The flux per pole is generated by five permanent magnets and hence strong flux-focusing effect can be obtained contributing to the improvement of force output, which is validated in [36]. In addition, the magnetization pattern of inner layer, i.e., axial magnetization not noly provides a strong flux-focusing effect but also guides the flux lines back to the airgap. In other words, the back iron of tradition quasi-Halbach array shown in Fig. 3.52a can be completely removed, as there is no flux loop in the inner side of the mover. Thus, the PMs of the proposed pattern can be mounted on the low-density nonferromagnetic tube to reduce mover mass greatly. The total length of dual-layer PMs (PM_{length}) is set to 6mm. In addition, the stator consist of 8-slot silicon steels and single-phase toroidal windings.

3.3.1.2 Working Principle

Due to the requirement of high-frequency and short-stroke reciprocating linear motion, the windings are energized by single-phase current and the current direction of adjacent coils is opposite to obtain high force output in short stroke, as shown in Fig. 3.53. The frequency of movement is directly related to the frequency of input current, which contribute to facilitate the control of this linear machine. According to this working principle, the force output of the proposed machine can be obtain by summing the electromagnetic forces on each coil.

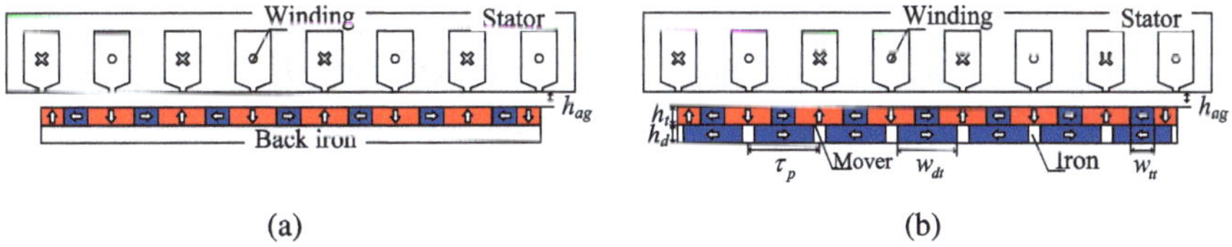

Fig. 3.52 Schematic structure. **a** Conventional quasi-Halbach pattern. **b** Proposed compound Halbach array

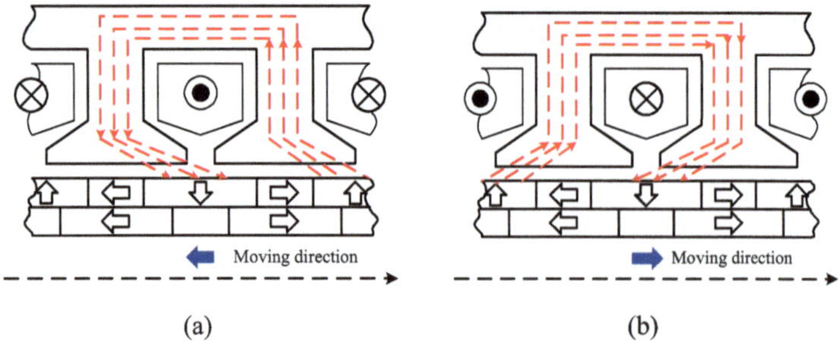

Fig. 3.53 Operation principle. **a** The movement of left direction. **b** The movement of right direction

3.3.2 Analysis of Mathematical Models

Although the dual-layer compound Halbach array provides strong flux-focusing and self-shielding effect, the complex structure also brings difficulties to the output force modeling. Thus, to obtain accurate mathematical model and hence the key parameters that are directly related with the electromagnetic performance, a novel virtual boundary-extension method is proposed in [36]. More explicitly, a mid air region is introduced into the field modeling as shown in Fig. 3.54. The flux density of five regions can be given by

$$\begin{aligned} \boldsymbol{B} &= \mu_0 \boldsymbol{H}, && \text{for region 1, 3, and 5} \\ \boldsymbol{B} &= \mu_0 \mu_r \boldsymbol{H} + \mu_0 \boldsymbol{M_0}, && \text{for region 2 and 4} \end{aligned} \tag{3.12}$$

where μ_0, $\boldsymbol{H}$ and $\boldsymbol{M_0}$ are the vacuum permeability, magnetic strength and remanent magnetization.

According to the Maxwell equations, $\nabla \cdot \boldsymbol{B} = 0$ and $\nabla \times \boldsymbol{B} \neq 0$, the flux density $\boldsymbol{B}$ can be written as the curl of magnetic vector potential $\boldsymbol{A}$, which satisfies

$$\nabla \times \boldsymbol{A} = \boldsymbol{B} \qquad \nabla \cdot \boldsymbol{A} = 0 \tag{3.13}$$

Equation $\nabla \cdot \boldsymbol{A} = 0$ is the Coulomb gauge used to simplify calculations. Besides, the magnetic field satisfies the constitutive relation between $\boldsymbol{B}$ and $\boldsymbol{H}$ and the Ampere's circuital law

Fig. 3.54 Solving region divisions in the mathematical models

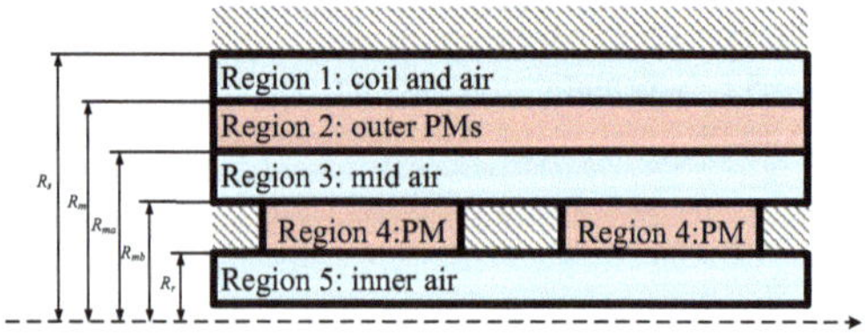

$$\boldsymbol{B} = \mu_0(\boldsymbol{H} + \boldsymbol{M}) \qquad \boldsymbol{J} = \nabla \times \boldsymbol{H} \tag{3.14}$$

where $\boldsymbol{J}$ is current density. Equations (2) and (3) are combined to give

$$-\nabla^2 \boldsymbol{A} = \mu_0(\boldsymbol{J} + \nabla \times \boldsymbol{M}) \tag{3.15}$$

where $\nabla^2 \boldsymbol{A} = (\delta_r, \delta_\theta, \delta_z)$, $\nabla \times \boldsymbol{M} = (N_r, N_\theta, N_z)$. The magnetic vector potential is determined by Eq. (4) and corresponding boundary conditions. According to the symmetry of flux density, the Eq. (4) is simplified as

$$\begin{aligned} &\frac{\partial}{\partial r}(\frac{1}{r}\frac{\partial}{\partial r}(rA_\theta)) = 0, && \text{for region1, 3, and 5} \\ &\frac{\partial}{\partial r}(\frac{1}{r}\frac{\partial}{\partial r}(rA_\theta)) = -\mu_0 N_\theta, && \text{for region2 and 4.} \end{aligned} \tag{3.16}$$

Then, according to Eq. (2), the flux density components satisfies

$$B_r = -\frac{\partial A_\theta}{\partial z} \qquad B_\theta = 0 \qquad B_z = \frac{1}{r}\frac{\partial (rA_\theta)}{\partial r} \tag{3.17}$$

The boundary conditions [36] are

$$\begin{aligned} &B_{z1}|_{r=R_s} = 0 \qquad B_{r5}|_{r=0} = 0 \\ &B_{r1}|_{r=R_m} = B_{r2}|_{r=R_m} \qquad H_{z1}|_{r=R_m} = H_{z2}|_{r=R_m} \\ &B_{r2}|_{r=R_{ma}} = B_{r3}|_{r=R_{ma}} \qquad H_{z2}|_{r=R_{ma}} = H_{z3}|_{r=R_{ma}} \\ &B_{r3}|_{r=R_{mb}\cup 0\le z\le \frac{w_{dt}}{2}} = B_{r4}|_{r=R_{mb}\cup 0\le z\le \frac{w_{dt}}{2}} \\ &H_{z3}|_{r=R_{mb}\cup 0\le z\le \frac{w_{dt}}{2}} = H_{z4}|_{r=R_{mb}\cup 0\le z\le \frac{w_{dt}}{2}} \\ &B_{r4}|_{r=R_{mb}\cup 0\le z\le \frac{w_{dt}}{2}} = B_{r5}|_{r=R_{mb}\cup 0\le z\le \frac{w_{dt}}{2}} \\ &H_{z4}|_{r=R_{mb}\cup 0\le z\le \frac{w_{dt}}{2}} = H_{z5}|_{r=R_{mb}\cup 0\le z\le \frac{w_{dt}}{2}} \\ &\int_{R_r}^{R_{mb}} rB_{z4}(r, \tfrac{w_{dt}}{2})dr = \int_{\frac{w_{dt}}{2}}^{\frac{\tau_p}{2}} (R_m B_{r3}(R_m, z) - R_r B_{r4}(R_r, z))dz \end{aligned} \tag{3.18}$$

where B_{z1} are the Z-axis component of flux density of region1. The B_{r1}, B_{r2}, B_{r3}, B_{r4}, B_{r5} are the R-axis component of flux density of every solving regions. Similarly, the H_{z1}, H_{z2}, H_{z3}, H_{z4}, H_{z5} are the Z-axis component of magnetization intensity of every solving regions. The slot effect should be taken into account for real airgap flux density distribution, and the details are expressed in [36]. Following that, the force output can be calculated by

$$F = \int_S (\boldsymbol{J} \times \boldsymbol{B})dS \tag{3.19}$$

It can be deduced from Eqs. (5), (6), (7) and (8) that the force output is related to the parameters including τ_p, W_{tt}, W_{dt}, R_s, R_m, R_{ma} and R_{mb}.

3.3.3 Characteristics of Output Performance

3.3.3.1 Force Output

Figure 3.55 shows the variation of average force output with the coefficients k_{outer} and k_{inner} under the case of $k_{length} = 0.1$. The k_{outer} is defined as the ratio of W_{tt} to τ_p, the coefficient k_{inner} is defined as the ratio of W_{dt} to τ_p and the coefficient k_{length} is defined as the ratio of W_{tt} to PM_{length}. The lager the k_{length}, the thicker the thickness of outer layer (quasi-Halbach array) is. It is found that the force output is maximum when $k_{outer} = 0.5$ and $k_{inner} = 0.7$. The k_{inner} should be large to obtain strong flux-focusing effect and hence large torque. However, a too large k_{inner} will lead to the decrease of the working point of the BH curve of the PMs due to mutual influence of the two axially magnetized PMs.

The result of the case of $k_{length} = 0.5$ is shown in Fig. 3.56. It can be seen that k_{outer} is almost the same and k_{inner} is a little smaller, when force output is the largest. The force output is larger than that of the case of $k_{length} = 0.1$. That is to say, the introduce of inner-layer pattern is beneficial for the improvement of thrust. In addition, the length of the axially magnetized PMs in the inner layer should be large to gain strong magnetomotive force, as in the case of $k_{length} = 0.1$.

Figure 3.57 shows the result of the case of $k_{length} = 0.9$. In other words, the compound Halbach array mainly consists of the quasi-Halbach array of outer layer under this case. It is found that the k_{inner} should be small to decrease the magnetic resistance. Also, the highest force output can be obtained when $k_{outer} = 0.5$. This is mainly due to the fact that the small and large k_{outer} results in sharp and flat top waveform of flux density in airgap, respectively. The largest force output under the case of $k_{length} = 0.9$ smaller than that of the case of $k_{length} = 0.5$. That is to say, there are optimal parameters, which are investigated in Sect. 5.

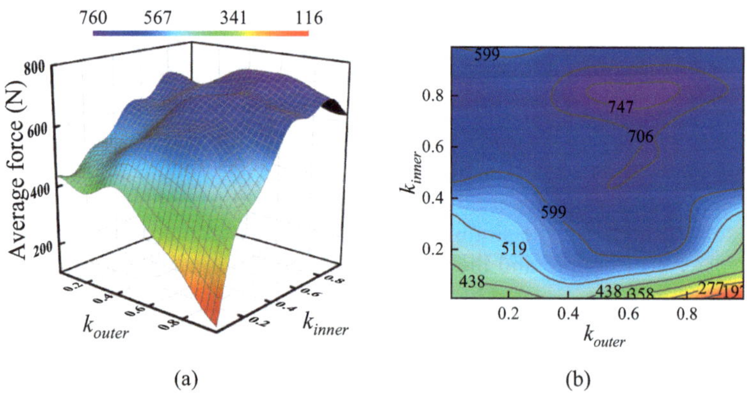

Fig. 3.55 Variation of force output with k_{inner} and k_{outer} under the case of $k_{length} = 0.1$. **a** Three-dimensional view. **b** Top view

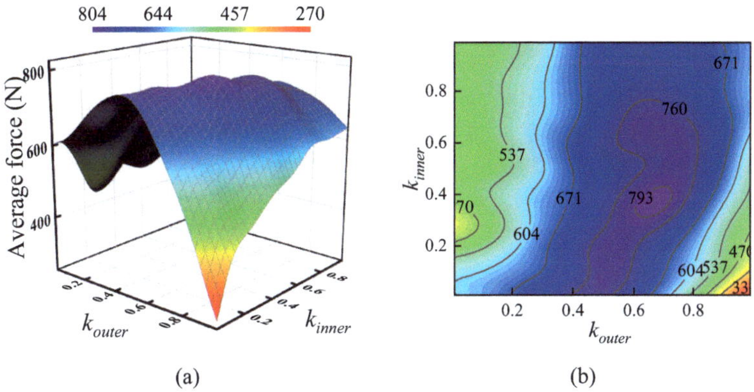

Fig. 3.56 The variation of force output with k_{inner} and k_{outer} under the case of $k_{length} = 0.5$. **a** Three-dimensional view. **b** Top view

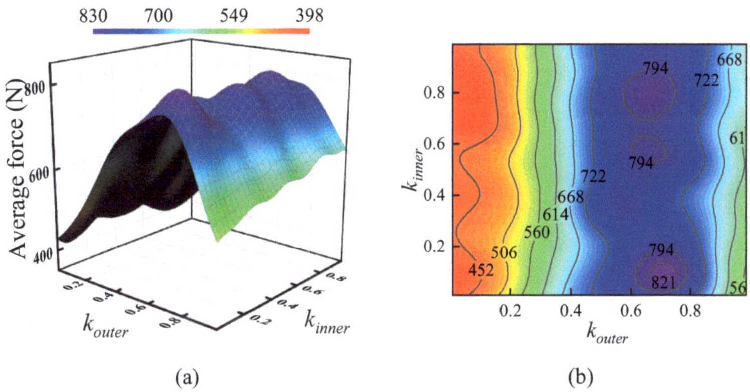

Fig. 3.57 The variation of force output with k_{inner} and k_{outer} under the case of $k_{length} = 0.9$. **a** Three-dimensional view. **b** Top view

3.3.3.2 Thrust Fluctuation

Figure 3.58 shows the variation of thrust fluctuation with the key parameters. With the increase of k_{length}, i.e., the increase of the thickness of outer quasi-Halbach array, the coefficient k_{outer} almost remains unchanged, and the value is about 0.2, at this time, the minimum thrust fluctuation can be obtained. Because the waveform of the airgap flux density is close to a flat-topped waveform, which contributes to reduce the variation of the magnetic field at the windings. In addition, the k_{inner} should be increased to obtain strong flux-focusing effect for high force output and hence small thrust fluctuation. It is worth noting that the thrust fluctuation of case of $k_{length} = 0.5$ is smaller than that of the other two cases.

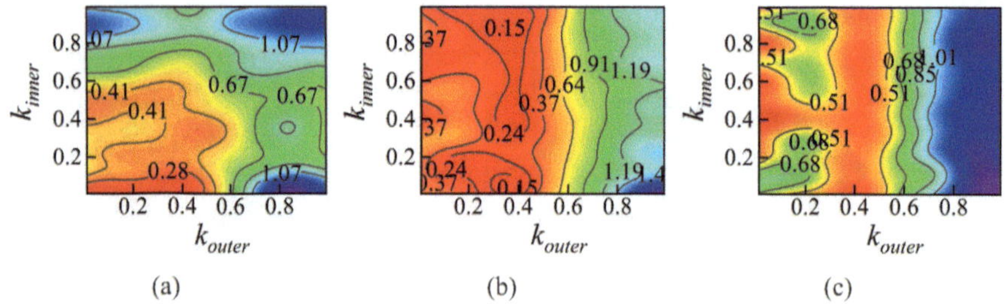

Fig. 3.58 Variation of thrust fluctuation with k_{inner} and k_{outer}. **a** $k_{length} = 0.1$. **b** $k_{length} = 0.5$. **c** $k_{length} = 0.9$

3.3.3.3 Energy Loss

Figure 3.59a shows the variation of stator core loss with the coefficients k_{inner} and k_{outer} under the case of $k_{length} = 0.5$. Larger k_{inner} and k_{outer} results in larger core loss, because with the increase of the coefficients k_{inner} and k_{outer}, the flux-focusing effect will be stronger. The results of core loss of the cases of $k_{length} = 0.1$ and $k_{length} = 0.9$ are similar to the case of $k_{length} = 0.5$. However, the high coefficient k_{inner} will leads to low core loss when k_{length} is large due to the high magnet resistance of the PMs in inner layer.

The variation of PM eddy loss with the key parameters are shown in Fig. 3.59b. It can be found that the PM eddy loss is small when the coefficients k_{outer} is large, which indicates that the W_{tt} with appropriate size can reduce PM eddy losses.

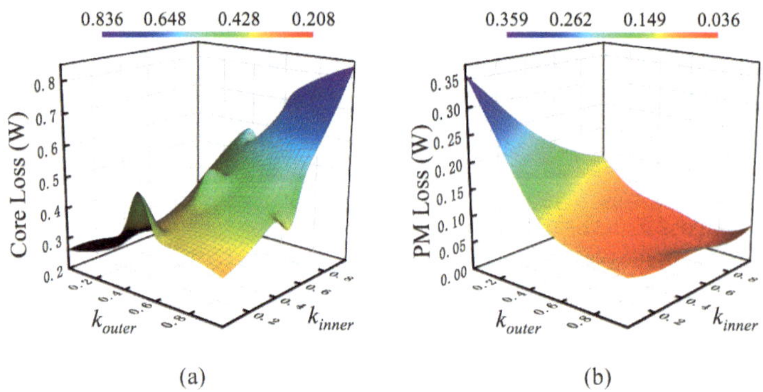

Fig. 3.59 The variation of energy loss with k_{inner} and k_{outer} under the case of $k_{length} = 0.5$. **a** Stator core loss. **b** PM eddy loss

3.3.4 Sizing Optimization

The key parameters, k_{inner}, k_{outer} and k_{length}, of compound Halbach array machine are globally optimized via multiobjective genetic algorithm. The constraints includes PM_{length} 6 mm, the tooth width 12 mm and airgap size 1 mm etc. For a better dynamic performance, the optimization objectives are the maximum average force and minimum thrust fluctuation. The results of sizing optimization are plotted in Fig. 3.60. The parameters of $k_{length} = 0.532$, $k_{inner} = 0.356$, $k_{outer} = 0.390$ are selected for the design of the compound Halbach array machine.

3.3.5 Comparison of Electromagnetic Characteristics

The electromagnetic characteristics including the flux distribution, force output and thrust fluctuation are compared with the conventional Halbach array. For a fair comparison, the thickness of conventional Halbach array is the same as that of the compound Halbach array. And the pole-arc coefficient k_{arc}, which is defined as the ratio of the width of axial magnetized PM to pole pitch, is optimized as shown in Fig. 3.61. For a stable force output, the coefficient $k_{arc} = 0.5$ is selected for the design of conventional Halbach array without back iron.

Figure 3.62a and b shows the open-circuit flux density of airgap and the inner side of the two arrays. It is found that the conventional Halbach array has a larger peak air-gap flux density, however, the compound Halbach array is more like a flat-

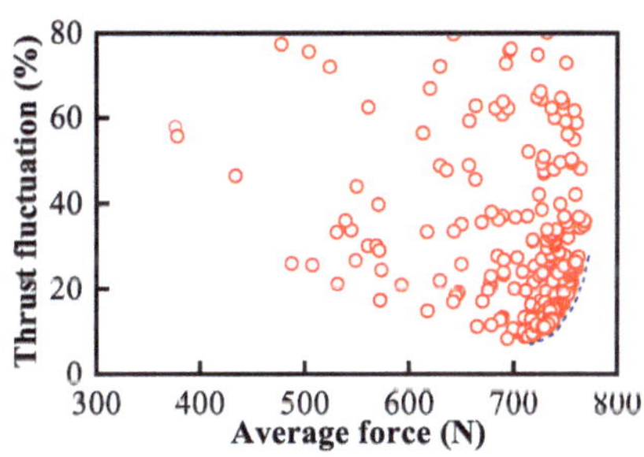

Fig. 3.60 Optimization result of average force and thrust fluctuation

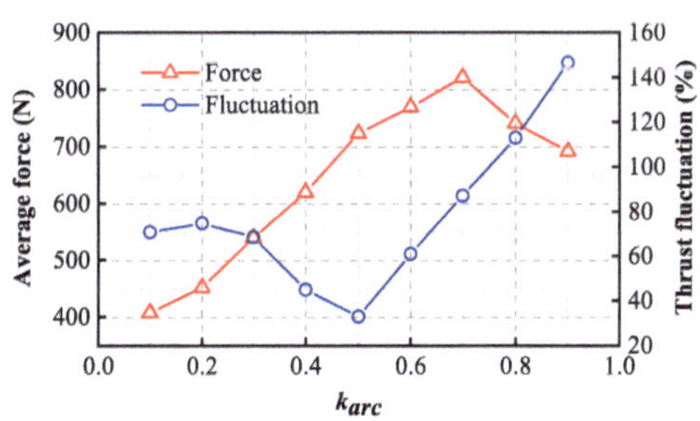

Fig. 3.61 Variation of force output and thrust fluctuation with k_{arc}

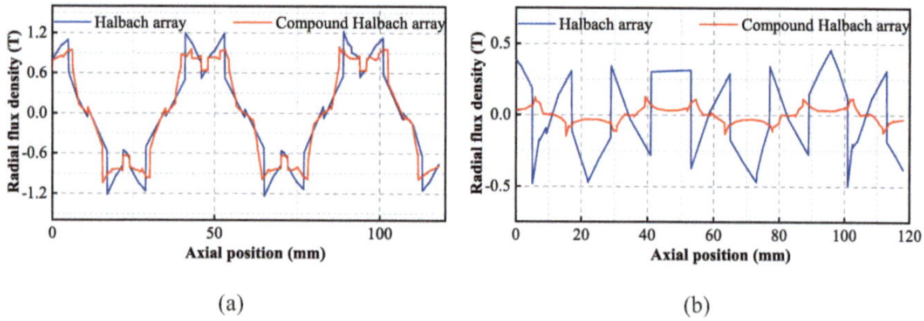

Fig. 3.62 Flux density distribution. **a** Airgap. **b** The inner side of arrays

Fig. 3.63 Variation of electromagnetic force with mover position

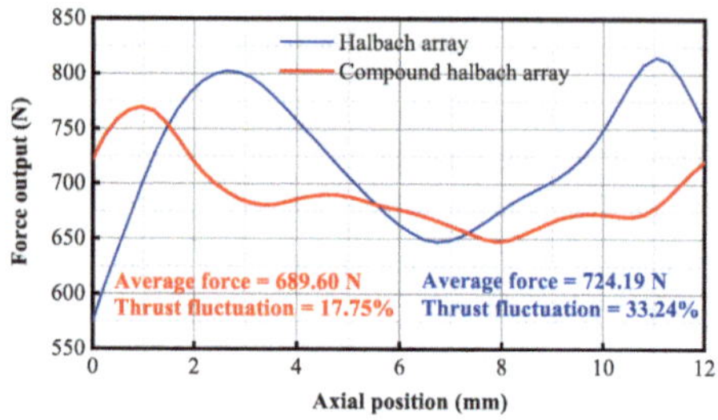

topped waveform which is beneficial to the generation of stable force output with the single-phase excited strategy. In addition, the compound Halbach array weakens the inner side flux density more than the conventional Halbach array, that is to say, it has a stronger self-shielding effect. Figure 3.63 shows the variation of electromagnetic force output with mover position using conventional Halbach and compound Halbach array. The compound Halbach array can provide more stable force output, i.e., smaller thrust fluctuation than conventional Halbach array. In addition, the proposed compound Halbach array has less PM usage with similar force output to conventional Halbach array because of the introduction of iron pole in the inner layer of the compound Halbach array.

3.4 Conclusion

In this chapter, the novel concentrated-flux-type PM machine with irregular magnets and iron poles is introduced for improving the torque output and utilization of rare earth PM material, and reducing the rotor mass, simultaneously. It has a few advantages as follows. (1) The flux density of airgap produced by neighboring quasi-trapezoid PMs through flux-focusing effect is enhanced, which helps to improve the torque output significantly, higher than conventional designs. (2) The CPMIP pattern with quasi-triangular iron poles provides good self-shielding effect, because there is no series branch in magnetic circuits. Thus the rotor yoke can be made from low-density materials, and the motion dynamics can be improved. The rotor mass of the CPMIP-1 is 49.11% of conventional radial and halbach array. And the rotor mass of the CPMIP-2 is 57.33% of conventional radial and halbach array. (3) The PM volume can be decreased, and PM utilization ratio, i.e., the ratio of average torque and PM volume, increases by 71% to conventional radial pattern and 64% to Halbach array, which helps to reduce the manufacturing cost much and maintain high torque output. In addition, compared with conventional CPM pattern, the proposed CPMIP-2 improves the average torque by 29.2%, PM utilization ratio by 12.3% and reduces the rotor yoke mass by 42.8%. Besides, the CPMIP patterns eliminates the unbalanced magnetic force that exists in the traditional CPM pattern.

Meanwhile, the novel compound Halbach array is also introduced for achieving high force output and low mover mass of the tubular linear oscillating machine simultaneously. The introduction of the inner layer pole pattern improves the self-shielding effect greatly. Thus, the conventional high-density back iron can be removed, which attributes to high dynamic performance. The coupling effect of these key design parameters on electromagnetic characteristics is investigated. Then, the multi-objective genetic algorithm is implemented for sizing optimization. Finally, the electromagnetic characteristics of the compound Halbach array are compared with conventional Halbach array. The results shows that the compound Halbach array has a stronger self-shielding effect than Halbach array.

References

1. S. Zheng, X. Zhu, L. Xu, Z. Xiang, L. Quan, and B. Yu, "Multi-objective optimization design of a multi-permanent-magnet motor considering magnet characteristic variation effects," IEEE Trans. Ind. Electron., vol. 69, no. 4, pp. 3428–3438, 2022.
2. Y. Zhou, R. Qu, D. Li, Y. Gao, and C. H. T. Lee, "Performance investigation and improvement of linear vernier permanent magnet motor for servo application," IEEE/ASME Trans. Mechatronics, pp. 1–13, 2023.
3. Z. Qiu, Y. Chen, X. Liu, L. Zhang, and H. Cheng, "Evaluation and com- parison of sideband harmonics and acoustic responses with continuous and discontinuous pwm strategies in pm synchronous motor for electric vehicles," Int. J. Hydromechatronics, vol. 5, no. 2, pp. 109–123, 2022.
4. P. Akiki, M. H. Hassan, M. Bensetti, P. Dessante, J.-C. Vannier, D. Prieto, and M. McClelland, "Multiphysics design of a V-shape IPM motor," IEEE Trans. Energy Convers., vol. 33, no. 3, pp. 1141–1153, 2018.
5. W. Geng, J. Wang, L. Li, and J. Guo, "Design and multiobjective optimization of a new flux-concentrating rotor combining halbach pm array and spoke-type ipm machine," IEEE/ASME Trans. Mechatronics, vol. 28, no. 1, pp. 257–266, 2023.
6. X. Ge, Z. Q. Zhu, J. Li, and J. Chen, "A spoke-type ipm machine with novel alternate airspace barriers and reduction of unipolar leakage flux by step-staggered rotor," IEEE Trans. Ind. Appl., vol. 52, no. 6, pp. 4789–4797, 2016.
7. L. Xu, W. Zhao, G. Liu, and C. Song, "Design optimization of a spoke- type permanent-magnet vernier machine for torque density and power factor improvement," IEEE Trans. Veh. Technol., vol. 68, no. 4, pp. 3446–3456, 2019.

8. S.-I. Kim, S. Park, T. Park, J. Cho, W. Kim, and S. Lim, "Investigation and experimental verification of a novel spoke-type ferrite-magnet motor for electric-vehicle traction drive applications," IEEE Trans. Ind. Electron., vol. 61, no. 10, pp. 5763–5770, 2014.
9. Q. Chen, G. Xu, F. Zhai, and G. Liu, "A novel spoke-type PM motor with auxiliary salient poles for low torque pulsation," IEEE Trans. Ind. Electron., vol. 67, no. 6, pp. 4762–4773, 2020.
10. M. M. Rahman, K.-T. Kim, and J. Hur, "Design and optimization of neodymium-free spoke-type motor with segmented wing-shaped pm," IEEE Trans. Magn., vol. 50, no. 2, pp. 865–868, 2014.
11. H.-J. Kim, D.-Y. Kim, and J.-P. Hong, "Structure of concentrated- flux-type interior permanent-magnet synchronous motors using ferrite permanent magnets," IEEE Trans. Magn., vol. 50, no. 11, pp. 1–4, 2014.
12. S. F. Rabbi and M. A. Rahman, "Critical criteria for successful synchro- nization of line-start IPM motors," IEEE J. Emerg. Sel. Topics Power Electron, vol. 2, no. 2, pp. 348–358, 2013.
13. X. Ge, Z. Zhu, J. Li, and J. Chen, "A spoke-type IPM machine with novel alternate airspace barriers and reduction of unipolar leakage flux by step-staggered rotor," IEEE Trans. Ind. Appl., vol. 52, no. 6, pp. 4789–4797, Nov.-Dec. 2016.
14. T. Wang and Z. Zhou, "A compact hydrostatic-driven electric generator: Design, prototype, and experiment," IEEE/ASME Trans. Mechatronics, vol. 21, no. 3, pp. 1612–1619, 2016.
15. S. Dwari and L. Parsa, "Design of halbach-array-based permanent- magnet motors with high acceleration," IEEE Trans. Ind. Electron., vol. 58, no. 9, pp. 3768–3775, 2011.
16. Z. Zhu, Z. Xia, and D. Howe, "Comparison of halbach magnetized brushless machines based on discrete magnet segments or a single ring magnet," IEEE Trans. Magn., vol. 38, no. 5, pp. 2997–2999, 2002.
17. X. Zhang, C. Zhang, J. Yu, P. Du, and L. Li, "Analytical model of magnetic field of a permanent magnet synchronous motor with a trapezoidal halbach permanent magnet array," IEEE Trans. Magn., vol. 55, no.7, pp. 1–5, 2019.
18. J. Qi, Z. Zhu, L. Yan, G. W. Jewell, C. Gan, Y. Ren, S. Brockway, and C. Hilton, "Effect of pole shaping on torque characteristics of consequent pole PM machines," IEEE Trans. Ind. Appl., 2022.
19. H. Yang, Z.-Q. Zhu, H. Lin, H. Li, and S. Lyu, "Analysis of consequent- pole flux reversal PM machine with biased flux modulation theory," IEEE Trans. Ind. Electron., vol. 67, no. 3, pp. 2107–2121, 2020.
20. S.-U. Chung, H.-J. Lee, B.-C. Woo, J.-W. Kim, J.-Y. Lee, S.-R. Moon, and S.-M. Hwang, "A feasibility study on a new doubly salient permanent magnet linear synchronous machine," IEEE Trans. Magn., vol. 46, no. 6, pp. 1572–1575, 2010.
21. Z. Wu and Z. Zhu, "Partitioned stator flux reversal machine with consequent-pole PM stator," IEEE Trans. Energy Convers., vol. 30, no. 4, pp. 1472–1482, 2015.
22. M. Onsal, Y. Demir, and M. Aydin, "A new nine-phase PM synchronous motor with consequent pole rotor for high-power traction applications," IEEE Trans. Magn., vol. 53, no. 11, pp. 1–6, 2017.
23. H. Hua, Z. Zhu, and H. Zhan, "Novel consequent-pole hybrid excited machine with separated excitation stator," IEEE Trans. Ind. Electron., vol. 63, no. 8, pp. 4718–4728, 2016.
24. D. Li, R. Qu, J. Li, and W. Xu, "Consequent-pole toroidal-winding outer- rotor Vernier PM machines," IEEE Trans. Ind. Appl., vol. 51, no. 6, pp. 4470–4481, 2015.
25. J. Li and K. Wang, "Analytical determination of optimal PM-arc ratio of consequent-pole permanent magnet machines," IEEE/ASME Trans. Mechatronics, vol. 23, no. 5, pp. 2168–2177, 2018.
26. J. Li, K. Wang, and C. Liu, "Torque improvement and cost reduction of PM machines with a dovetailed consequent-pole rotor," IEEE Trans. Energy Convers., vol. 33, no. 4, pp. 1628–1640, 2018.
27. J. Li, K. Wang, F. Li, S. Zhu, and C. Liu, "Elimination of even-order harmonics and unipolar leakage flux in consequent-pole PM machines by employing ns-iron-sn-iron rotor," IEEE Trans. Ind. Electron., vol. 66, no. 3, pp. 1736–1747, 2019.

Chapter 4
Multi-Objective Stator/Rotor Structural Optimization Methods

Abstract Conventionally, the designers of electrical machines assume the employment of regular shaped magnets such as rectangular, sector or tiled-shapes etc. The geometry of magnet poles is fixed, and the design purpose is mainly to optimize the dimensional values. However, these regular shapes employed may not be the best choice, unavoidably limiting the final design result, and compromising the output performance of the electrical machines. An alternative scheme is to conduct topology optimization for rotor poles. By this method, the geometrical shape of magnets along with the iron core is no longer fixed, but automatically obtained through the optimal distribution of materials such as iron, magnet and air. It helps to achieve better optimization results with less shape constrains. Therefore, this article proposes a multi-material topology optimization method based on variable density representation for surface-mounted permanent-magnet synchronous motor (PMSM) to obtain an optimal rotor pole pattern, and thus improve the output performance of electrical machines, especially the torque density which is crucial in aerospace industry. The computational speed is increased by improving the objective function and the arbitrary volume constraints become clear by combining size and topology optimization. To validate the design method, theoretical, numerical and experimental studies are conducted. First, the operating principle, design variables, constrains, optimization objective and procedure of the method are introduced in detail. Then the optimization method is implemented into the rotor design of one surface-mounted electric machine, and the result is presented and discussed. Subsequently, the magnetic flux density and torque output of electric machine with the optimized rotor poles is compared with conventional design, and it shows that the former can achieve better output performance relatively. Finally, one research prototype of the electrical machine is developed, and experiments are carried out on torque output and back-EMF. The experimental results validate the optimization method. In the other hand, when designing a motor, the input current density is used to calculate the output torque. When studying the influence of structural parameters on motor performance, the output torque always changes, resulting in the inability to immediately obtain the optimal design at rated torque. Therefore, an optimization method for fixed output torque is proposed to address the issue of the inability to obtain the optimal solution for the stator design of a permanent magnet synchronous motor directly under rated

P. Xiang et al., *Rotor Pole Pattern Topology Technology for Permanent Magnet Electric Machine*, https://doi.org/10.1007/978-981-95-2018-3_4

conditions. This method simplifies the calculation process and is proved high precision. By conducting sensitivity analysis on the stator teeth of the stator core, the impact of stator teeth width on the motor output performance is obtained, and the optimal solutions for torque and efficiency are obtained. Comparing the two optimal solutions, it is found that the performance of the efficiency optimal solution is better. Furthermore, it is found that the stator material only works below the knee point of its BH curve when the efficiency is optimal. Therefore, the optimal solution of the motor is not directly related to the knee point of the stator, and using the method described in this chapter can achieve better performance.

4.1 Introduction

Permanent magnet synchronous motors (PMSM) with surface-mounted magnet rotor can achieve large torque density, high efficiency, and fast dynamic response [1, 2], and thus have wide applications in aerospace technology [3–5]. In practical engineering applications, the designers of PMSM usually select the number of slots and poles firstly, and then determine the stator bore through empirical computation [6]. The magneto-motive force of the surface mounted PMSM mainly comes from the permanent magnets [7–9], so the optimization design of the rotor impacts the output performance of the electrical machine significantly, especially the torque density which is crucial in aerospace industry.

One typical optimization method is the parameter sweep or sensitivity analysis approach. For example, the influence of pole arc coefficient on torque characteristics was studied by Yang et al. [10]. The pole arc coefficient is optimized through parameter sweep approach to achieve the lowest cogging torque. Similarly, as shown in Fig. 4.1, Zheng et al. [11] employed an eccentric magnetic pole and optimized the eccentric distance through the same approach. However, in these studies, only one parameter is swept to obtain the optimal solution, which may constrains the performance improvement more or less. Furthermore, although the cogging torque is reduced, the average torque output is also compromised.

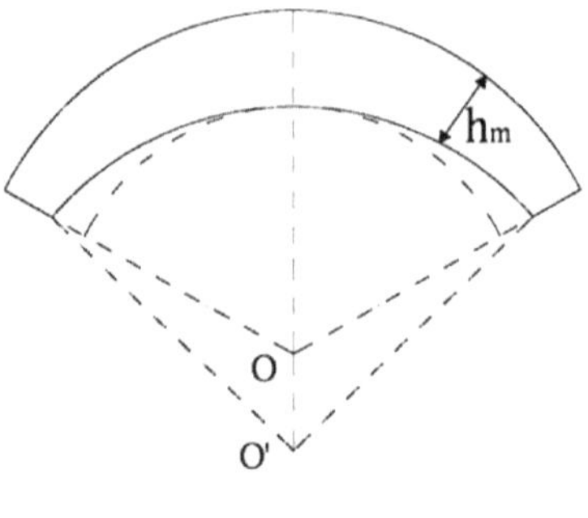

Fig. 4.1 Shape of the eccentric magnetic pole in [11]

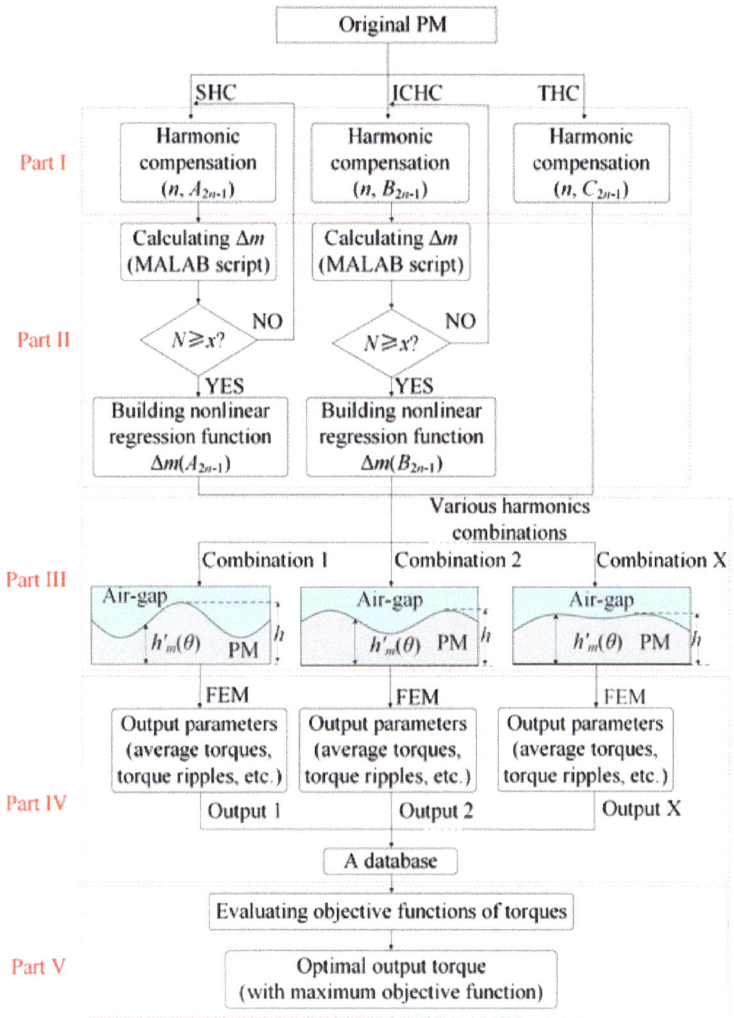

Fig. 4.2 Optimization procedure of proposed SHC, ICHC, and THC methods in [12]

To overcome the above problems, the parameter sweep approach with multiple variables have been proposed by researchers. For example, the harmonic compensation pole is proposed to reduce the cogging torque and maintain the average torque of surface-mounted PMSM by Zeng [12] and Zhu et al. [13–15]. The corresponding rotor profiles are shown in Figs. 4.2, 4.3 and 4.4. The sinusoidal arc shaping with the third order harmonic term and the inverse cosine shaping with the third order harmonic term are optimized simultaneously. Other methods have been

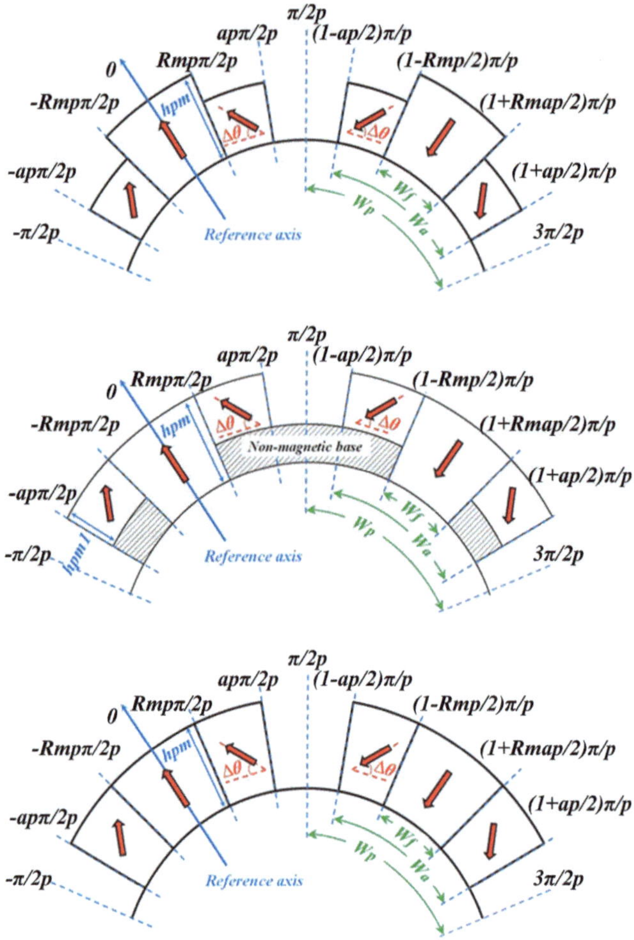

Fig. 4.3 Design parameters for Hat- and T-type rotor magnet poles [13]

proposed to achieve similar effects of harmonic compensation [16, 17], as shown in Figs. 4.5 and 4.6. These optimization methods only considers the influence of magnets and neglects the impact of iron, which may constrains the performance improvement of the electrical machine. Therefore, Xiang et al. [18] proposed one high-performance concentrated-flux-type PMSM with irregular magnets and iron poles. The parameter sweep approach with multiple variables is employed to optimize the magnet and the iron together, and thus to improve the torque output and reduce the rotor mass.

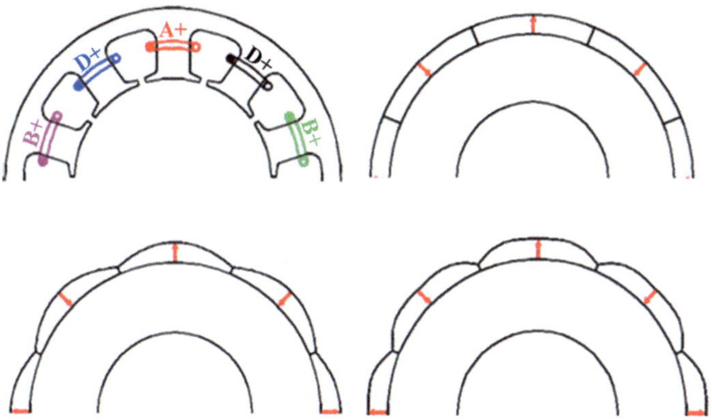

Fig. 4.4 10-slot/8-pole SPM machines with different PM shapes in [15]

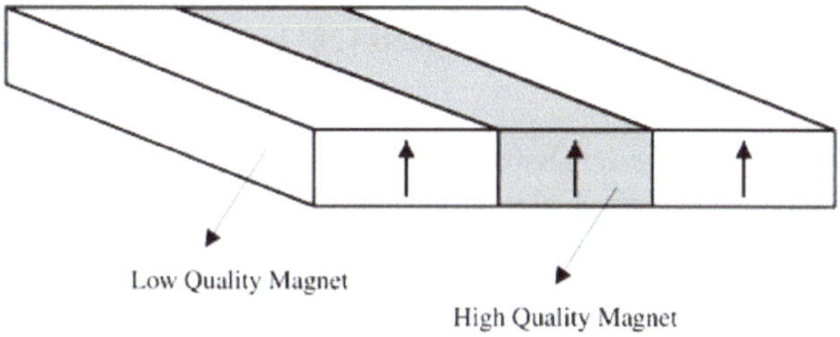

Fig. 4.5 Proposed modular PM pole with different material quality in [16]

Fig. 4.6 Segmented pole imitation by virtual PM blocks in [17]

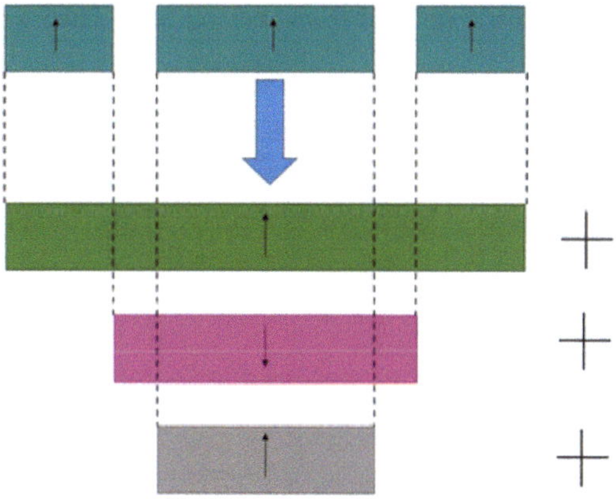

However, so far, most design optimization methods for PMSM assume that the magnet and iron geometry is fixed or expressed by parametric equations, and only the dimensional values or parameters are optimized. As shown in Fig. 4.7, these studies typically focus on size optimization or shape optimization in detailed structural design [19]. It can only change the dimensional of a hole, but cannot transform one hole to more holes. Thus, they may overlook other topologies that could potentially achieve better performance. These topological shapes can be achieved through topology optimization, which modifies the distribution of materials [20], as shown in Fig. 4.8. To deal with multi-materials optimization in design of magnetic devices, the Optimized material Distribution (OMD) method is presented by Dyck et al. [21]. To achieve the maximum magnetic field strength in a specific region, the magnetization direction of the permanent magnet is also optimized [22–24], as shown in Figs. 4.9 and 4.10. This technique is also employed in the design of interior PMSM [25, 26], as shown in Fig. 4.11.

Nevertheless, research on topology optimization for surface-mounted PMSM is still limited. Although studies addressing problems related to asymmetric magnetic poles [27] and non-linear permanent magnets [28] by Cherriere et al., there are still some challenges:

1. *Low Computational Speed*: Density-based topology optimization relies on mesh of finite element analysis. Calculating the average output torque of the motor takes a considerable amount of time.
2. *Arbitrary volume constraint*: For current topology optimization methods of surface-mounted PMSM, the proportion of material volume is determined arbitrarily, which may compromise the output performance in a certain degree.

Therefore, the purpose of this article is to propose one multi-material topology optimization method based on variable density representation. The computational

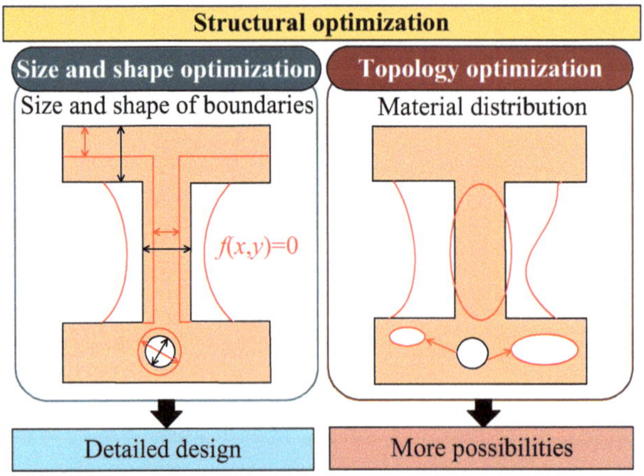

Fig. 4.7 The relationship between topology, shape and size optimization

cost of topology optimization is reduced by improving the objective function and the arbitrary volume constraints become clear by combining size optimization and topology optimization. The rest is organized as follows. In Section II, the operating principle of the multi-material topology optimization method based on variable density representation is introduced, and the design variables, the constraints and the optimization objectives are presented. In Section III, the proposed optimization method is implemented into one surface-mounted radial magnetized PMSM, and the optimization result is discussed. In Section IV, the performances of the PMSM by the proposed optimization method are compared with that of conventional optimization method. Finally, in Section V, one research prototype of the electrical machine with the optimized structure has been developed, and experiments are conducted to validate the output performance.

On the other hand, improving the efficiency of electric permanent magnet synchronous motor (PMSM) is an important issue in the field of electrical engineering [40–43]. There are many methods to achieve this goal, but one of the most effective ways is to use higher-grade electromagnetic materials and improve manufacturing processes. For example, using high permeability and low-loss core materials [44–46], high conductivity windings, increasing stacking factor, and improving slot filling factor [47–50], among others. However, the improvement of material performance means an increase in cost, and achieving efficiency improvement solely through material selection is limited. What truly reflects design level is maximizing the utilization of existing materials and achieving further efficiency improvement. This method includes optimizing the design of the motor and its components (such as the stator and rotor) to reduce losses and increase power output. By doing so, greater efficiency improvement can be achieved than solely through material selection.

The motor structure has a significant impact on motor efficiency, and appropriate structural dimensions can achieve efficient utilization of various materials [51–55]. To get an appropriate geometry of a high efficiency motor, many scholars have done a lot of research. As shown in Fig. 4.12, Boztas et al. [56] employed a multi-objective

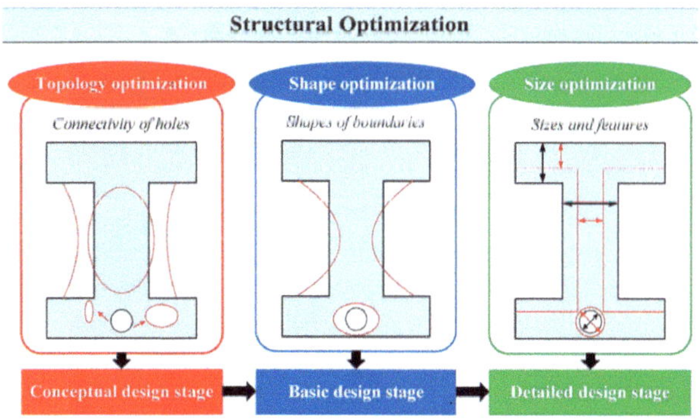

Fig. 4.8 Structural optimization process in [20]

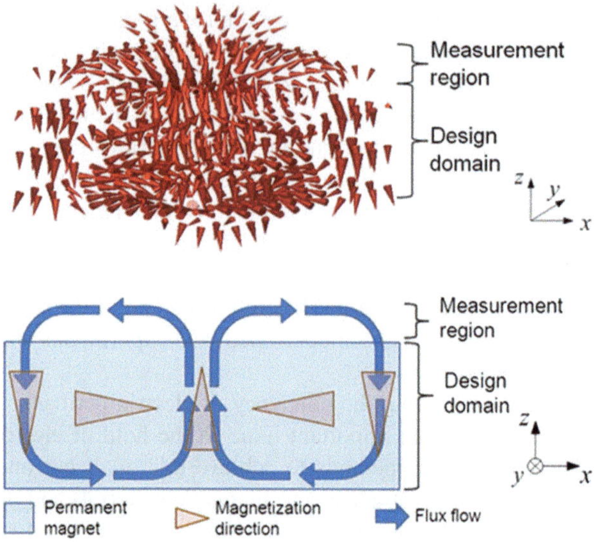

Fig. 4.9 The research model in [22]

Fig. 4.10 Fabrication results in [24]

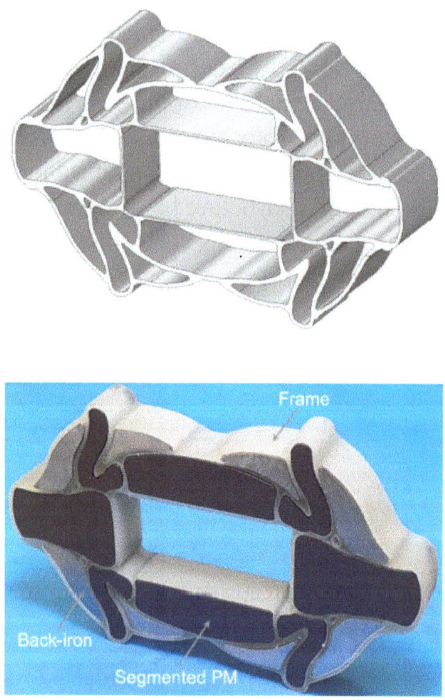

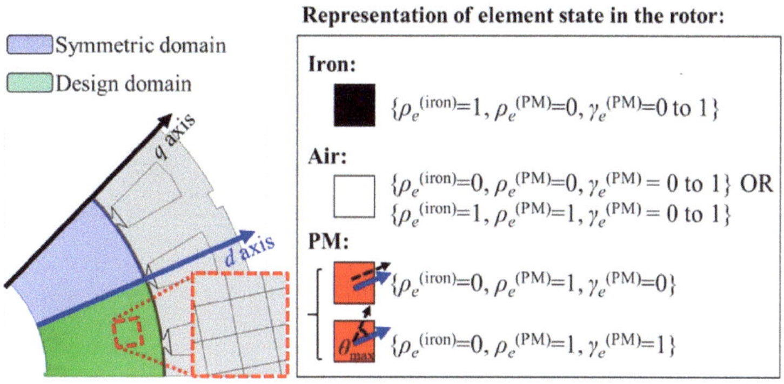

Fig. 4.11 Three-material states determined in the electromagnetic topology optimization in [26]

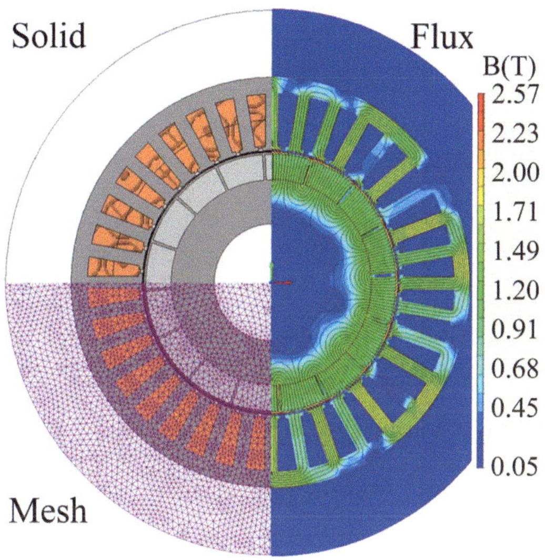

Fig. 4.12 Mesh and solid view and flux density distribution of the optimized motor in [56]

genetic algorithm to optimize the geometric structure of an electric motor, resulting in a high power density motor. As shown in Fig. 4.13, Weiduo Zhao et al. [57] employ a multi physics optimization program based on multi-objective genetic algorithm to achieve a compromise solution in electromagnetic, mechanical, and thermal aspects. As shown in Fig. 4.14, Sun et al. [58] design a PMSM for a campus patrol electric vehicle by different optimization processes.

Although these methods have achieved high performance through some optimization algorithms, they cannot interpret the impact of some structural parameters on performance. Taking the stator tooth width as an example, a stator tooth that is too wide will result in a small slot area, making it difficult to embed many enameled wires, leading to a reduction in torque. On the other hand, a stator tooth that is too narrow will cause tooth saturation, increasing iron losses and reducing torque. Some litera-

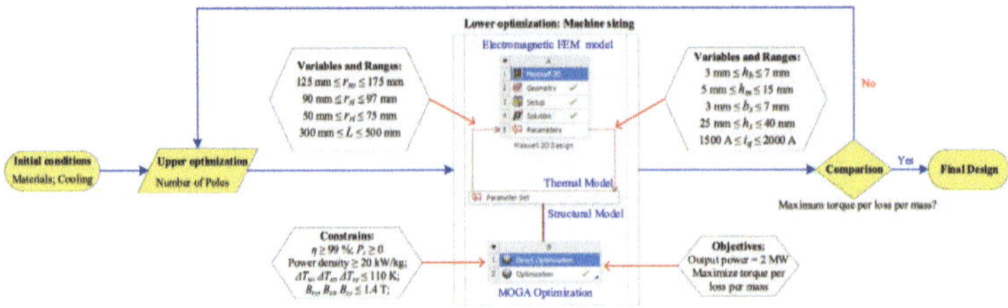

Fig. 4.13 Illustration of the optimization process in [57]

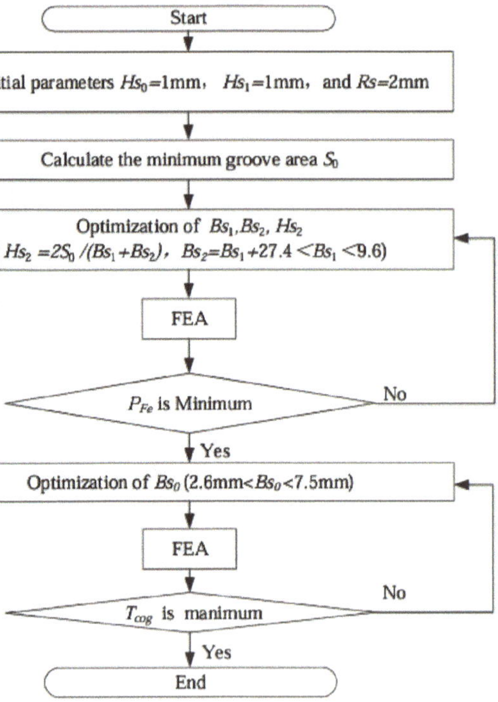

Fig. 4.14 Stator optimization procedure in [58]

ture mentions designing the stator tooth to the knee point of the soft magnetic material saturation, but this conclusion is subjective, as soft magnetic materials often do not have an obvious knee point. This design method is too dependent on the experience of the designer. In addition, these optimized motors can achieve good performance, but they cannot guarantee optimal efficiency under rated operating conditions. Especially for motors used in fields such as pumps and fans, their rated operating point is often a specific value, and currently there is no method for optimizing design for rated operating conditions.

To solve this problem, this chapter proposes a stator tooth optimization design method that constrains the output operating condition and uses electromagnetic driving efficiency as the optimization objective through analyzing the electromagnetic driving efficiency.

4.2 Interpolation Functions and Field Methods in Electric Machine Optimization

4.2.1 Work Principle

4.2.1.1 Multi-Material Optimization Based on SIMP

To achieve multi-material optimization within the design domain, i.e. the materials of the domain maybe Neodymium PM, iron or air, the design domain is divided into finite elements. the constitutive equations within these elements can be written as

$$\mathbf{B} = \mu_0 \mu_\mathrm{r} \mathbf{H} + \|\mathbf{B}_\mathrm{r}\| \, \mathbf{e} \tag{4.1}$$

where **B** is the magnetic flux density, **H** is the magnetic field intensity, μ_0 is the magnetic permeability of vacuum which is a constant of $4\pi \times 10^{-7}$ T · m/A, **e** is the magnetization direction of permanent magnets, μ_r and $\|\mathbf{B}_\mathrm{r}\|$ are the relative permeability and remanence that determine the material properties in elements. As shown in Table I, μ_{iron} and B_{rem} are the permeability of iron and remanence of NdFeB, respectively.

In order to control the relative magnetic permeability μ_r and remanence $\|\mathbf{B}_\mathrm{r}\|$ within the elements, two density variables, $\alpha_c \in [0, 1]$ and $\beta_c \in [0, 1]$, are introduced. This method relies on a finite element grid, which may result in a checkerboard pattern or jagged boundariess. Additionally, since alpha and beta are continuous variables, while it is desired for them to be either 0 or 1, intermediate variables, or grayscale variables, can introduce material properties that lack practical significance. Therefore, the intermediate variables need to be minimized as much as possible. To achieve this, the Helmholtz filtering method [13] is initially applied to filter the density variables, i.e.

$$\alpha_\mathrm{f} = R\nabla^2 \alpha_\mathrm{f} + \alpha_\mathrm{c}, \beta_\mathrm{f} = R\nabla^2 \beta_\mathrm{f} + \beta_\mathrm{c} \tag{4.2}$$

where α_f and β_f is the density variables after filtering, is the filtering radius, which is related to the grid size. Then the hyperbolic tangent projection method [14] is adopted, i.e.

$$\alpha = \frac{\tanh(k_\alpha(\alpha_f - \alpha_0)) + \tanh(k_\alpha \alpha_0)}{\tanh(k_\alpha(1 - \alpha_0)) + \tanh(k_\alpha \alpha_0)} \tag{4.3}$$

$$\beta = \frac{\tanh(k_\beta(\beta_f - \beta_0)) + \tanh(k_\beta \beta_0)}{\tanh(k_\beta(1 - \beta_0)) + \tanh(k_\beta \beta_0)} \tag{4.4}$$

where α_0 and β_0 are midpoint of the projections, k_α and k_β are slope of the projections. Ultimately, clear and smooth material boundariess are attained. Finally, appropriate interpolation functions are designed to control the relative magnetic permeability and remanence as detailed in the next section.

4.2.1.2 Interpolation Functions

There are four combinations of alpha and beta, namely (0,0), (0,1), (1,0), and (1,1). However, there can only be three types of material properties, as shown in Table 4.1. The state of $\mu_r = \mu_{iron}$ & $\|\mathbf{B}_r\| = B_{rem}$ have no meaning in reality, so they have no meaning in reality, so they should not exist. Therefore, a very natural interpolation method is

$$\|\mathbf{B}_r\| = B_{rem}\alpha^m(1 - \beta)^m \tag{4.5}$$

$$\mu_r = 1 + [\mu_{iron} - 1]\beta^n(1 - \alpha)^n \tag{4.6}$$

where m and n are the penalty factors.

The NdFeB used as the hard magnetic material here has a recoil magnetic permeability close to 1 and is almost linearly changing. The magnetic permeability of soft magnetic materials varies with the change of magnetic field. When the external magnetic field is applied to a certain extent, the soft magnetic material reaches a saturation state. Therefore, the interpolation function of soft magnetic materials has a significant impact on magnetic permeability. In order to compare the influence of different interpolation functions on magnetic permeability, the penalty factors and are taken as 1, and alpha is taken as 0, i.e.

$$\mu_r = 1 + (\mu_{iron} - 1)\,\beta \tag{4.7}$$

Table 4.1 Material attribute values and representation

Relative permeability μ_r	Remanence $\|\mathbf{B}_r\|$	The materials represented
1	0	Air
1	B_{rem}	NdFeB
μ_{iron}	0	Iron
μ_{iron}	B_{rem}	Non-existent

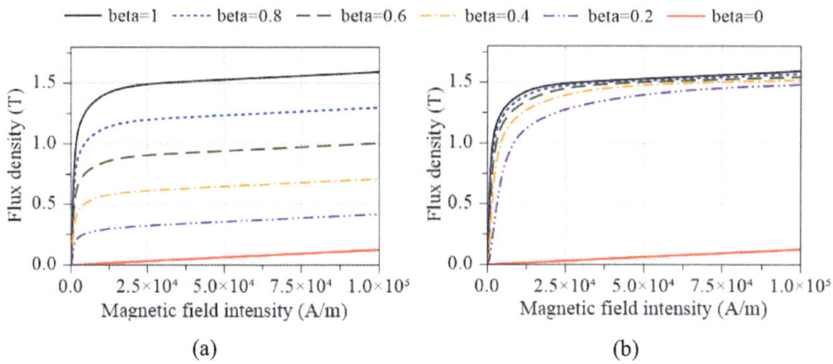

Fig. 4.15 Permeability interpolated **B**-**H** curve based on **H** and **B**

where μ_{iron} can be calculated based flux density **B** or magnetic field intensity **H** according to the **B**-**H** curve tested from the iron specimen. The final output **B**-**H** curve of the interpolated relative permeability as shown in Fig. 4.15, where (a) represents the influence of density variables on the **B**-**H** curve calculated based on magnetic field intensity, (b) represents the influence of density variables on the BH curve calculated based on magnetic flux density.

The magnetization permeability interpolation method based on magnetic field intensity exhibits a gradual increase in saturated magnetic flux density as the density variable increases and the magnetic field intensity at the knee is similar. This interpolation holds physical significance, representing a scenario where numerous magnetic flux conductors are parallelly connected within the grid unit. With an increase in the cross-sectional area of these conductors, more magnetic flux passes through the grid, leading to an augmentation in saturated magnetic flux density. On the other hand, the magnetization permeability interpolation method based on magnetic flux density lacks clear physical significance. Moreover, the density variable within the range of 0 to 1, sharing most characteristics with soft magnetic materials, results in a significant contribution of intermediate variables to the magnetic field. This, in turn, generates numerous intermediate variables during the optimization process.

Another commonly used interpolation function involves interpolating the magnetic reluctance [11], i.e., the derivative of magnetic permeability. That is:

$$\frac{1}{\mu_{\text{r}}} = 1 + \left(\frac{1}{\mu_{\text{iron}}} - 1\right)\beta \tag{4.8}$$

The final output B-H curve of the interpolated relative permeability as shown in Fig. 4.16, where (a) represents the influence of density variables on the BH curve calculated based on magnetic field intensity, (b) represents the influence of density variables on the **B**-**H** curve calculated based on magnetic flux density.

The magnetic reluctivity interpolation method based on magnetic field intensity yields a **B**-**H** curve that appears irregular, and this interpolation method lacks a clearly

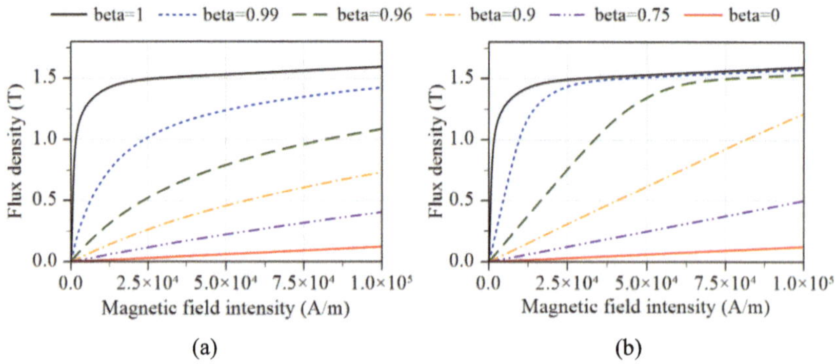

Fig. 4.16 Reluctivity interpolated **B-H** curve based on **H** and **B**

defined physical significance. On the other hand, the magnetic permeability interpolation method based on magnetic flux density allows for the clear observation that, with an increase in the density variable, the magnetic flux density at the saturation knee point remains relatively constant. This interpolation holds physical significance, representing a scenario where numerous magnetic flux conductors are connected in series within the grid unit. As the total length of these conductors increases, the applied magnetic field intensity required for the conductors to reach saturation gradually decreases. The intermediate variables in both interpolation methods contribute minimally to the magnetic field, facilitating the formation of clear material boundaries.

4.2.1.3 Magnetic Field Calculation Methods

The essence of magnetic field calculations appears to involve solving a system of differential equations, and there are two methods for this. One method involves using the magnetic vector potential as the dependent variable in magnetic field calculations, and this method is widely applicable. The other method involves using the magnetic scalar potential as the dependent variable in magnetic field calculations, and this method can only address scenarios where there is no current within the magnetic field. Dividing both sides of 4.1 by μ and taking the curl yields:

$$\frac{1}{\mu_0\mu_{\mathrm{r}}}\nabla \times \mathbf{B} = \nabla \times \mathbf{H} + \frac{1}{\mu_0\mu_{\mathrm{r}}}\left\|\mathbf{B}_{\mathrm{r}}\right\| \nabla \times \mathbf{e} \tag{4.9}$$

Due to the divergence-free nature of magnetic flux density, it can be expressed as the curl of some field. Therefore, the magnetic vector potential **A** is introduced for electromagnetic field calculations, so the flux density can be calculated as

$$\mathbf{B} = \nabla \times \mathbf{A} \tag{4.10}$$

Then substituting Maxwell's equation $\mathbf{J} = \nabla \times \mathbf{H}$ in magnetic field and 4.10 into formula 4.9 obtains:

$$\frac{1}{\mu_0 \mu_r}[\nabla(\nabla \cdot \mathbf{A}) - \nabla^2 \mathbf{A}] = \mathbf{J} + \frac{1}{\mu_0 \mu_r} \|\mathbf{B}_r\| \nabla \times \mathbf{e} \tag{4.11}$$

Upon introducing the Coulomb gauge $\nabla \cdot \mathbf{A} = 0$, Eq. 4.11 can be transformed into the following Poisson equation as

$$\frac{1}{\mu_0 \mu_r} \nabla^2 \mathbf{A} = -\frac{1}{\mu_0 \mu_r} \|\mathbf{B}_r\| \nabla \times \mathbf{e} - \mathbf{J} \tag{4.12}$$

After solving for vector potential **A**, the magnetic flux density is computed using 4.10, and subsequently, the magnetic field intensity is calculated based on material properties 4.1 as

$$\mathbf{H} = \frac{1}{\mu_r \mu_0}(\mathbf{B} - \|\mathbf{B}_r\| \mathbf{e}) \tag{4.13}$$

This process leads to the determination of the entire electromagnetic field.

When $\mathbf{J} = 0$, i.e. **H** is a irrotational field, it can be expressed as the gradient of a scalar potential. Therefore, the magnetic scalar potential can be introduced to calculate the electromagnetic field, and the magnetic field intensity can be computed as

$$\mathbf{H} = -\nabla V_m \tag{4.14}$$

Substituting 4.14 into 4.1 obtains

$$\mathbf{B} = -\mu_0 \mu_r \nabla V_m + \|\mathbf{B}_r\| \mathbf{e} \tag{4.15}$$

As **B** is a divergence-free field, $\nabla \cdot \mathbf{B} = 0$. Therefore, taking the divergence of both sides of 4.15, we can obtain the following Poisson equation as

$$\mu_0 \mu_r \nabla^2 V_m = \|\mathbf{B}_r\| \nabla \cdot \mathbf{e} \tag{4.16}$$

After solving for scalar potential V_m, the magnetic field intensity is computed using 4.14, and subsequently, the magnetic flux density is calculated based on material properties 4.1. This process leads to the determination of the entire electromagnetic field.

4.2.2 Comparison of Interpolation Functions

A multi-material topology optimization is conducted for the surface-mounted permanent magnet motor, focusing on the magnetic pole which plays a crucial role in the performance of the motor. For example, Xiang et al. proposed an innovative radial-flux surface-mounted PM machine with quasi-cylindrical pole pattern. His interesting design can help to reduce the cogging torque and the torque ripple of electrical machines significantly. Thus, to further elaborate on the impact of interpolation functions on optimization results, interpolation functions are employed for magnetic permeability based on magnetic flux density, magnetic reluctance, and magnetic permeability based on magnetic field intensity, respectively.

4.2.2.1 Topology Optimization of a PMSM Pole

The optimization model is illustrated in Fig. 4.17, where the design domain can consist of three materials: permanent magnet, back iron, and air. Within the air gap, there is a line used to extract the radial air gap magnetic flux density. For a surface-mounted permanent magnet synchronous motor, the fundamental magnetic flux density is directly proportional to the motor's output torque, thus the optimization objective is to maximize the fundamental radial air gap magnetic flux density along this line, denoted as

$$\text{Maximize } B_{r1} = \frac{2p}{\pi} \int_{-\frac{\pi}{2p}}^{\frac{\pi}{2p}} B_r(\theta)\cos(p\theta)d\theta \tag{4.17}$$

where p is polar distance, θ is the position of the probe curve in polar coordinates, and B_{r1} is the fundamental radial air-gap flux density.

The constraints include the material proportions of the permanent magnet and back iron, represented as:

$$\frac{\int_{\Omega} \alpha ds}{A_{\Omega}} \leq \rho_{\text{PM}}^*, \frac{\int_{\Omega} \beta ds}{A_{\Omega}} \leq \rho_{\text{Iron}}^* \tag{4.18}$$

where A_{Ω} is the area of the design domain, ρ_{PM}^* and ρ_{Iron}^* represent the material ratios of PM and iron in design domain, respectively. Here, the material ratios are arbitrarily chosen as $\rho_{\text{PM}}^* = 0.28$ & $\rho_{\text{Iron}}^* = 0.14$.

The commercial software COMSOL v6.2 is used for analysis. Globally convergent method of moving asymptotes (GCMMA) [15] along with adjoint method is used as the optimization solver for sensitivity analysis and updating density variables.

4.2.2.2 Permeability Interpolation Based on Flux Density

Permeability interpolation based on flux density as 4.5 is adopted to optimize the pole shown in Fig. 4.17. In this optimization, m is set as 5. Due to the nonlinear properties of ferromagnetic material, the value of n has significant impact on the optimization results. As shown in Fig. 4.18, red area represents the permanent magnet and blue area represents the back iron. When $n = 3$, although there are clear boundariess among the three materials, many intermediate variables appear in the iron region, which may result in inaccurate calculation of the material proportion and deviation from optimal design; when $n = 9$, the effect of the magnetic permeability decreases, and the calculation is unstable; when $n = 15$, the influence of the magnetic permeability is even worse, causing inability to form a concentrated material area and optimization failure. These phenomena represent the specific manifestation of the numerical instability associated with the magnetic permeability interpolation based on magnetic flux density.

To solve this problem, multiple penalty coefficients are employed during each iteration and the average of the computed objective functions is utilized as the new objective function. As a result, the final optimization can be described as

$$\text{Maximize}\frac{1}{8}\sum_{i=1}^{8} B_{r1}|_{n=2i-1} \tag{4.19}$$

With the objective function in 4.19, the clear boundariess of the three materials can be obtained, especially between the iron and the air as shown in Fig. 4.18d.

Another method to address this issue is referred to as the continuation method, which involves progressively increasing the penalty factor with each iteration step.

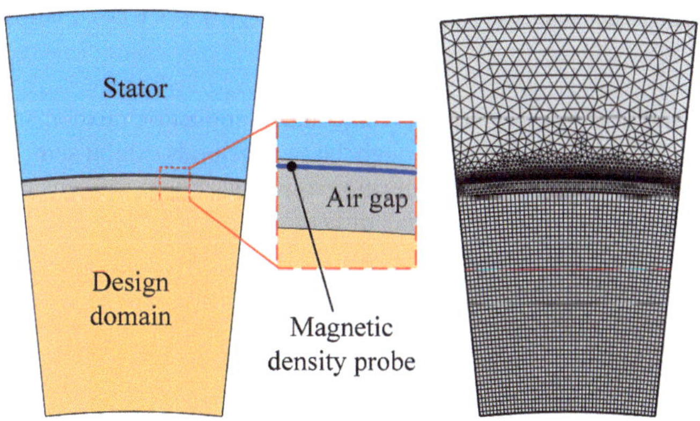

Fig. 4.17 Figure 3. Optimization model of rotor magnetic pole topology

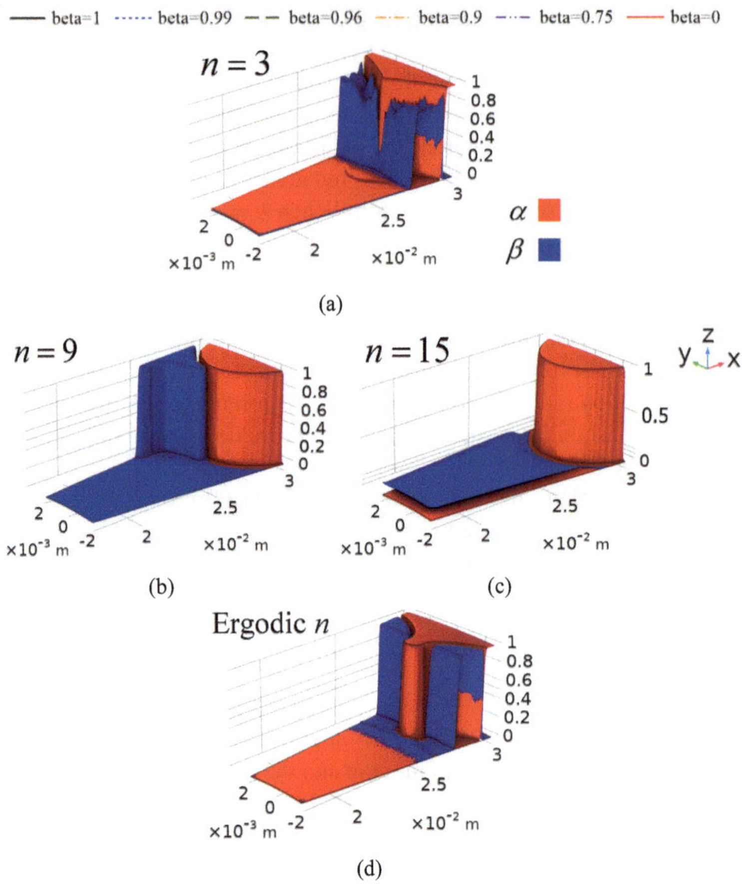

Fig. 4.18 Optimization results with different penalty strategy. **a** n=3. **b** n=9. **c** n=15. **d** Ergodic n

The results obtained using average method and continuation method are illustrated in Fig. 4.19. The results obtained through the two methods exhibit some differences, with the continuity method's optimization appearing more refined. In the results from the averaging method, the boundariess of the permanent magnet and back iron are concave, while the continuity method yields convex boundariess. However, the optimized results from both methods are very similar, featuring a small convexity in the permanent magnet portion. This is because, with this interpolation method, even when the density variable is small, the material properties remain close to those when the density variable is 1. Even in the boundary regions of soft magnetic materials, positive effects on the magnetic field are observed. Theoretically, the small protrusion in the center contributes minimally to the torque output. In practice, this protruding shape increases manufacturing complexity and is susceptible to break-

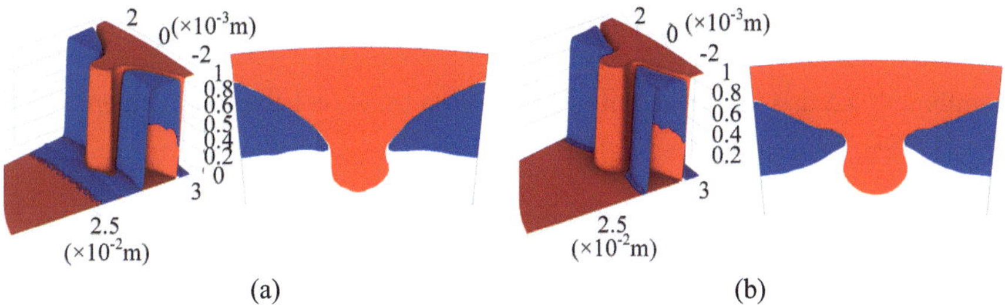

Fig. 4.19 Optimization results with average method and continuation method. **a** Average method. **b** Continuation method

age. Therefore, the shape obtained through topology optimization using magnetic permeability interpolation based on magnetic flux density is not practical.

4.2.3 Match of Interpolation and Calculation Methods

The current mainstream viewpoint favors the stability of using magnetic reluctance interpolation over magnetic permeability interpolation, and as a result, the latter is gradually being phased out. This section asserts that the stability of the optimization process is related to the dependent variable in the computational method, and interpolation functions should involve multiplication with the dependent variable rather than division. This is because when the interpolation function appears in the denominator and approaches zero, it can lead to the magnitude of the computed field vector tending toward infinity, causing computational instability.

For instance, in solid structural topology optimization, the ratio of stress to strain is the Young's modulus, with strain as the dependent variable. Interpolating the Young's modulus based on strain yields stable results. As mentioned earlier, magnetic reluctance interpolation based on magnetic flux density has a clear physical significance. Using the magnetic vector potential as the dependent variable allows the calculation of magnetic flux density first, followed by the interpolation of magnetic reluctance multiplied by magnetic flux density to obtain magnetic field intensity. If permeability interpolation is used, it requires dividing by magnetic flux density in the interpolation function, resulting in unstable computational outcomes.

Similarly, magnetic permeability interpolation based on magnetic field intensity also has a clear physical meaning. If the magnetic scalar potential is used as the dependent variable for electromagnetic field calculation, obtaining magnetic field intensity first and then calculating permeability multiplied by magnetic field intensity should yield stable results.

4.2.3.1 Reluctivity Interpolation Based on Flux Density

Combining 4.5 and 4.8, reluctivity interpolation function based on flux density can be obtained as

$$\|\mathbf{B}_\mathrm{r}\| = B_{\mathrm{rem}}\alpha^m(1-\beta)^m \tag{4.20}$$

$$\frac{1}{\mu_\mathrm{r}} = 1 + [\frac{1}{\mu_{\mathrm{iron}}} - 1]\beta^n(1-\alpha)^n \tag{4.21}$$

Figure 4.20 illustrates the iterative process of topology optimization: under the initial conditions where alpha and beta are both set to 0.5, the first iteration occurs, and alpha and beta are reduced to satisfy the material proportion constraints. Starting from the second step, alpha gathers at the top of the magnetic pole, forming the permanent magnet region. Subsequently, at both ends of the permanent magnet region, beta begins to accumulate, forming the soft magnetic region. By the 7th iteration, the permanent magnet and soft magnetic material regions are fully formed, and the boundariess become increasingly clear. Iterating beyond the 25th step, the shape almost no longer changes, and the objective function tends to stabilize.

As shown in Fig. 4.21, validation of the optimized magnetic pole shape reveals a uniform distribution of magnetic flux density in the soft magnetic material, approximately at 1.47T, which corresponds to the knee point of the BH curve for the soft magnetic material. This indicates that the soft magnetic material is maximally utilized.

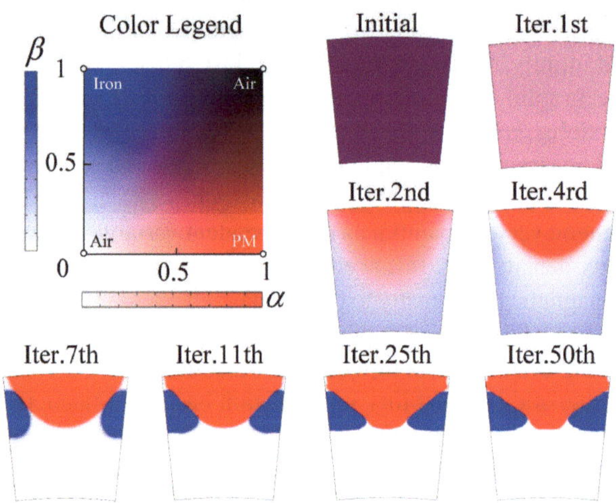

Fig. 4.20 Optimization results with different penalty strategy

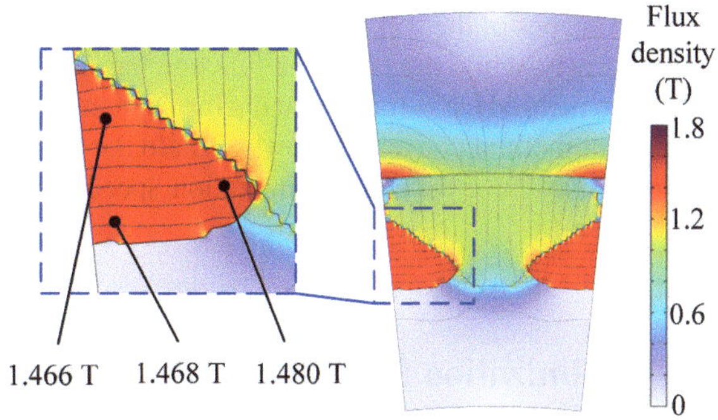

Fig. 4.21 Optimization results with different penalty strategy

4.2.3.2 Permeability Interpolation Based on Field Intensity

To validate the applicability of permeability interpolation in electromagnetic field calculations using magnetic scalar potential, a topology optimization is conducted for this problem using the magnetic permeability interpolation based on magnetic field intensity. The optimization results are shown in Fig. 4.22.

Initial state is the same as the topology optimization with reluctivity interpolation based on flux density. However, the difference lies in that, in the second step of this method, the permanent magnet and soft magnetic materials are distributed at two extremes. Subsequently, the distribution of soft magnetic material gradually approaches the magnetic pole. By the 14th iteration, clear boundariess have already

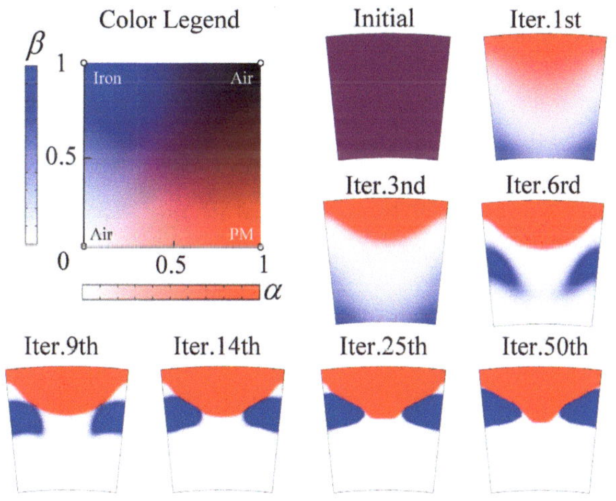

Fig. 4.22 Optimization results with different penalty strategy

formed. Afterward, the shapes of the soft magnetic material and the permanent magnet continue to move slowly, while the objective function has already stabilized.

Comparing these two methods, it is found that the former converges more quickly. Although the subsequent method converges slowly, due to the fewer computational degrees of freedom, it may have a better optimization speed advantage in three-dimensional optimization.

4.3 Efficiency Optimization of PMSM Stator Tooth Design

4.3.1 Electromagnetic Model

The permanent magnet synchronous motor (PMSM) is chosen as the subject of this study due to its advantages of high efficiency and high power density. The motor's topology is shown in 4.23, with a surface-mounted rotor and two-layer winding whose dimensions and structure have been determined. The stator's outer diameter, inner diameter, and air gap have also been determined and are listed in Table 4.2. The material assignment of each component is shown in Table 4.3. The motor is designed to operate at a rated speed of 1800 r/min and a power of 2.4 kW. The optimization of the stator tooth width is now being carried out to achieve the highest efficiency at the rated operating point while keeping the existing material properties and dimensions.

The present study utilizes motor-CAD for electromagnetic calculations. The traditional electromagnetic calculation process involves calculating the number of turns based on the back electromotive force coefficient and designing the diameter and number of parallel wires of the electromagnetic wire based on the actual slot fill factor, considering the slot area. Subsequently, given the peak or effective value of the current in the winding and the rated speed of the motor, the calculation results

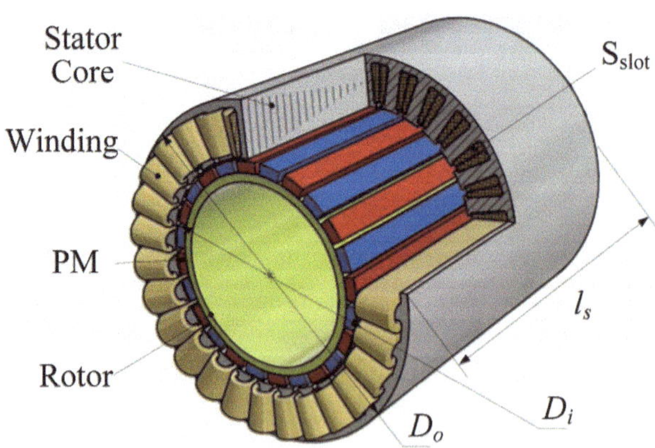

Fig. 4.23 Geometric structure of permanent magnet synchronous motor

Table 4.2 Geometric structural parameters of motor

Geometric parameter	Value	Geometric parameter	Value
Slot number	27	Pole number	24
Stator diameter	86 mm	Air gap	0.6 mm
Stator bore	58 mm	Magnet thickness	2.8 mm
Slot depth	12 mm	Magnet Arc[ED]	146 deg
Tooth tip depth	1 mm	Tooth tip angle	10 deg
Slot opening	1.7 mm	Shaft diameter	46.8 mm
Stator length	100 mm	End winding overhang	2 times

Table 4.3 Material assignment of each component

Component	Material
Stator lamination	VACOFLUX 50 0.35 strip
Winding	86 Copper (Pure)
Rotor lamination	M250-35A
Magnet	12 N42UH

are obtained through finite element analysis. The motor's electromagnetic efficiency η can then be calculated using the input speed and obtained output torque, copper loss, and iron loss, as follows:

$$\eta = \frac{nT_e}{nT_e + P_{Cu} + P_{Fe}} \tag{4.22}$$

where n is the rated speed of the motor shaft; T_e is the electromagnetic torque of the motor; P_{Cu} is copper loss and P_{Fe} is iron loss.

However, the method presents the following issues (as shown in Fig. 4.24):

1. The torque obtained in each calculation step is not the rated torque, and iterative updating of the driving current is required to obtain the rated torque under rated conditions.
2. When the coupling parameters of the motor change, the winding parameters, such as the number of turns, parallel branch number, and number of parallel wires, need to be modified simultaneously, resulting in a cumbersome process.

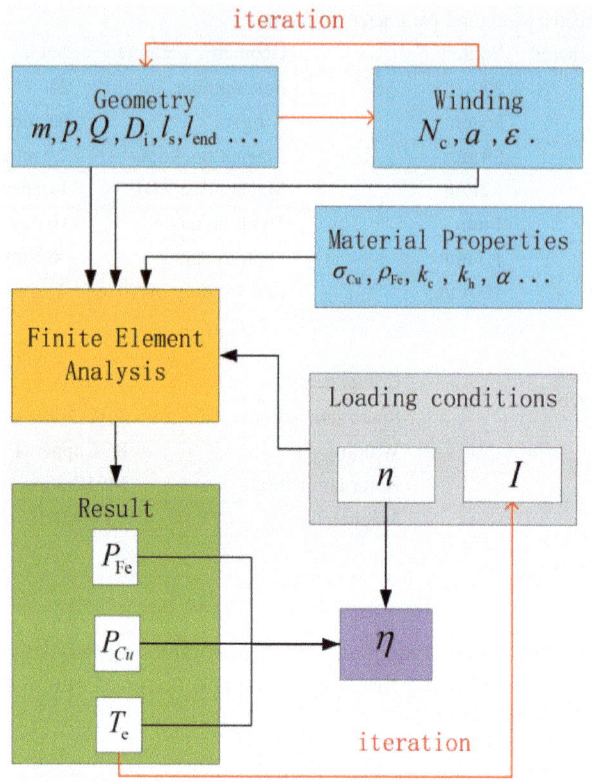

Fig. 4.24 Improved design and optimization process of a PMSM

4.3.2 *Calculation Model Improvement*

Assuming that the motor has m phase, and the number of parallel branches is a, turns per phase N can be calculated as

$$N = \frac{QN_c}{ma} \tag{4.23}$$

where Q is the slot number and N_c is the turns in a single coil.

While the length of each electromagnetic wire l_{whole} can be calculated as

$$l_{\text{whole}} = 2N(l_s + l_{\text{end}}) \tag{4.24}$$

where l_s is the length of stator core and l_{end} is the length of end winding.

If there are ε strands in one turn and the gross slot fill of copper in slot is k_f, the cross-sectional area of a single stranded wire is

$$S_{Cu} = \frac{S_{\text{slot}} k_f}{2\varepsilon N_c} \tag{4.25}$$

where S_{slot} is slot area.

Therefore, the phase resistance R can be expressed as

$$R = \frac{l_{\text{whole}}}{a\varepsilon\sigma S_{Cu}} \tag{4.26}$$

where σ is the conductivity of copper.

The relationship between phase current I and the density of current J in the slot is

$$I = Ja\varepsilon S_{Cu} \tag{4.27}$$

So the copper loss of the motor P_{Cu} can be directly calculated using current density as

$$P_{Cu} = mI^2R = \frac{1}{\sigma}Q(l_s + l_{\text{end}})S_{\text{slot}}k_f J^2 \tag{4.28}$$

It can be observed that, if the geometry determined, the copper loss in a motor is directly proportional to the square of the current density. The number of turns in the slot, number of parallel branches, and number of winding strands do not have any effect on the calculation of copper loss. Therefore, by directly utilizing the current density in the slot as an input, it becomes unnecessary to calculate the winding parameters. The only required input in this case is the slot fill factor of copper.

The fundamental magnetomotive force (MMF) of the stator winding is

$$F_1 = \frac{\sqrt{2}mNk_{w1}}{\pi p}I \tag{4.29}$$

where p is the number of pole pairs, k_{w1} is winding factor.

The magnetic flux of each pole is

$$\Phi_1 = \frac{B_{av}\pi D_i l_s}{2p} \tag{4.30}$$

where B_{av} is average magnetic density within the air gap.

The electromagnetic torque from a magnetic field perspective is calculated by [59]

$$T_e = \frac{\pi}{2}p^2 F_1 \Phi_1 \sin\delta_1 \tag{4.31}$$

where δ_1 is approximately equal to the angle between the fundamental wave of the stator MMF and the rotor MMF. In order to maximize the torque, $\delta_1 = 90°$, i.e. $\sin\delta_1 = 1$ should be made.

As a consequence, the electromagnetic torque can also be obtained through current density as

$$T_e = \frac{\sqrt{2}\pi}{8} Q k_{wl} B_{av} D_i l_s S_{slot} k_f J \tag{4.32}$$

It can be seen that in addition to constants and coefficients in the equation, there are two types of variables that determine the relationship between torque and current density. One is the geometric parameter, and the other is the air gap magnetic density B_{av}. Although B_{av} is affected by current density, due to the much smaller magnetic field generated by the armature compared to the permanent magnet, when the geometric size is determined, B_{av} is almost determined, and the electromagnetic torque is proportional to the current density.

In addition, the relationship between copper loss and torque can also be obtained by

$$P_{Cu} = \frac{32(l_s + l_{end})T_e^2}{\pi^2 k_{w1}^2 k_f \sigma Q D_1^2 l_s^2 S_{slot} B_{av}^2} \tag{4.33}$$

If T_{rate} is the rated torque, the copper loss under rated torque P'_{Cu} can be calculated by

$$P'_{Cu} = \frac{T_{\text{rate}}^2}{T_e^2} P_{Cu} \tag{4.34}$$

Due to the negligible relationship between other losses and current density, the electromagnetic efficiency η under rated operating conditions can be directly calculated by

$$\eta_{\text{rate}} = \frac{n T_{\text{rate}}}{n T_{\text{rate}} + P_{Cu} T_{\text{rate}}^2 / T_e^2 + P_{Fe}} \tag{4.35}$$

As a result, the electromagnetic model is improved as shown in Fig. 4.25. By adopting an improved calculation model, the electromagnetic efficiency under the target working condition can be obtained without considering the winding parameters in the initial design stage, and without adjusting the current density after structural changes. This eliminates the iterative calculation process and greatly saves calculation time.

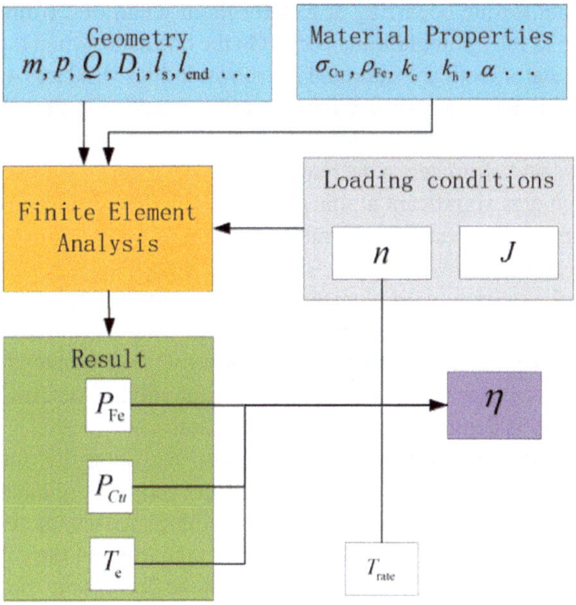

Fig. 4.25 Improved design and optimization process of a PMSM

4.3.3 *Sensiticity Analysis of Tooth Width*

To validate the aforementioned theory, the electromagnetic design of the aforementioned motor was optimized by varying the stator tooth width to achieve maximum efficiency at rated conditions. The method for defining the line current employed the line current density (12A/mm^2) and the winding copper fill factor was fixed at 0.4. A parameterized scan of stator tooth widths was performed within the range of 2 mm to 5 mm, resulting in a series of calculated results.

From Fig. 4.26, it can be observed that with an increasing tooth width and constant current density, the output torque first increases and then decreases, with the

Fig. 4.26 Dependence of output torque on tooth width

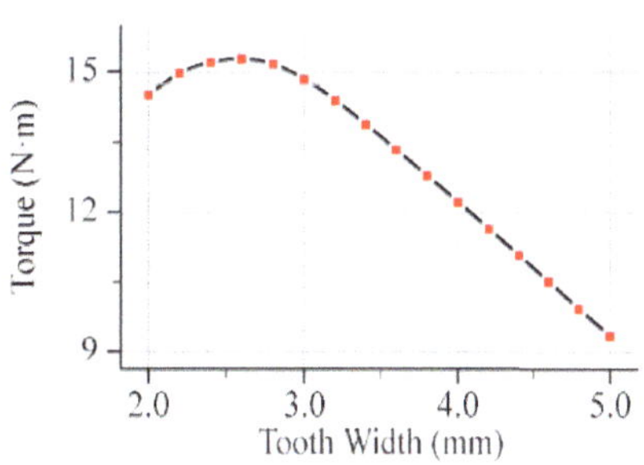

maximum output torque occurring at a stator tooth width of 2.6 mm. This point is later referred to as the optimal torque point. On the one hand, when the tooth width is very small, the stator teeth are saturated (as shown in Figs. 4.30 and 4.31), leading to a significant reduction in the air gap magnetic flux density. On the other hand, when the stator teeth are too large, the slot area decreases, resulting in a reduction in the number of turns and ultimately a decrease in torque. However, as shown in Fig. 4.27, the iron loss remains significant at this point. With an increasing stator tooth width, the slot area gradually decreases, resulting in a corresponding decrease in copper loss.

Besides, as the saturation problem at the teeth is improved (as shown in Figs. 4.30 and 4.31), the iron loss rapidly decreases, followed by a slow decrease. As the torque and losses decrease, the motor efficiency first increases and then tends to stabilize as the tooth width increases (Fig. 4.29), making it impossible to determine whether the optimal efficiency is achieved at the operating point.

The rated torque of this motor is 12.733 Nm. By using this torque to correct the calculation results, the loss of the fixed output torque is obtained as shown in Fig. 4.28. It can be observed that as the tooth width increases, the copper loss first decreases and then increases, which is opposite to the trend of the calculated output torque. This means that in order to compensate for the decrease in torque, the current density must be increased, leading to an increase in copper loss. Comparing the efficiency curves obtained by the two methods (Fig. 4.29), it can be concluded that the efficiency curve obtained by the improved calculation method has a peak point,

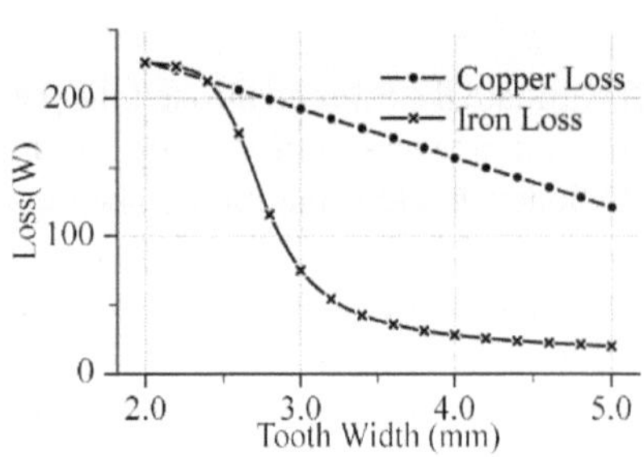

Fig. 4.27 Loss changes with tooth width in fixed current density

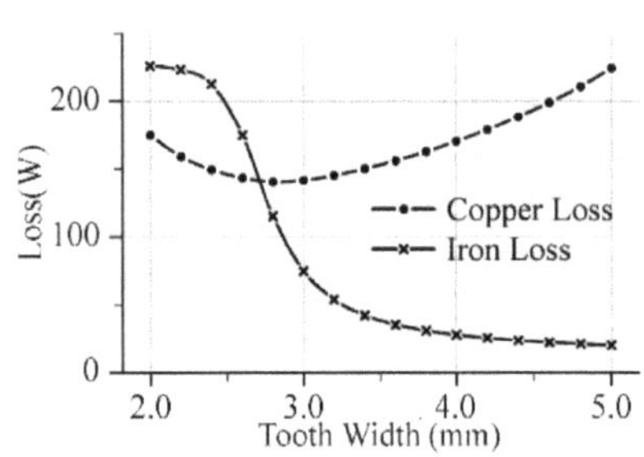

Fig. 4.28 Loss changes with tooth width in fixed rated torque

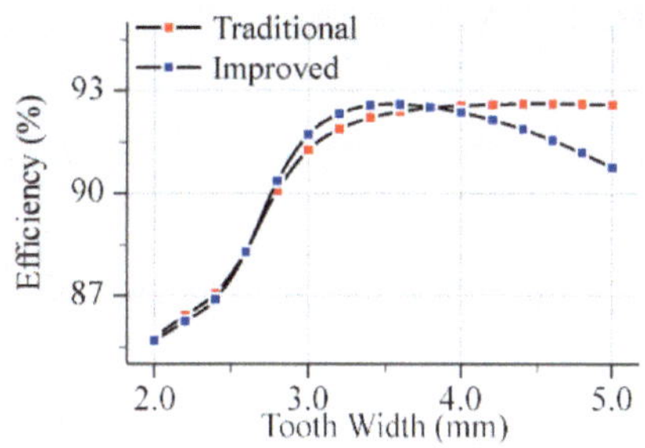

Fig. 4.29 Comparison of two calculation results

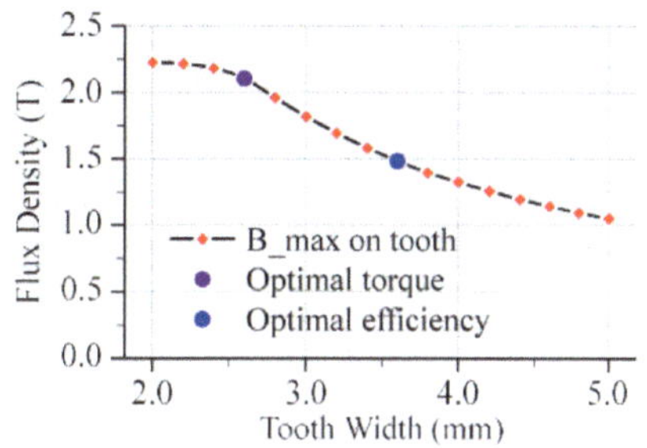

Fig. 4.30 Dependence of average flux in tooth on tooth width

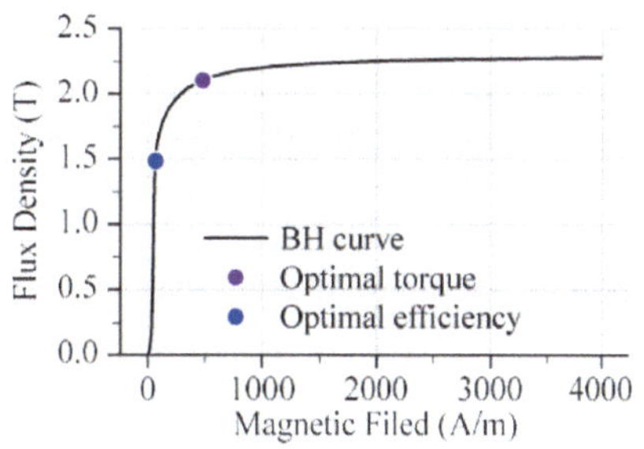

Fig. 4.31 Working points of material in tooth

with the motor achieving maximum efficiency at a stator tooth width of 3.6 mm under rated operating conditions. This point is later referred to as the optimal efficiency point.

4.3.4 Comparison of Optimization Results

After analyzing the sensitivity of the teeth, two optimal points were obtained, namely the torque optimal point (tooth width = 2.6 mm) and the efficiency optimal point (tooth width = 3.6 mm). In order to achieve the rated output torque, the current density of the two points should be set to 10.003 A/mm^2 and 11.453 A/mm^2 respectively. The magnetic field diagram is shown in Fig. 4.32, and the output powers obtained are 2405.8 W and 2398.8 W, respectively. The error with the proposed rated power is only

Fig. 4.32 Magnetic field distribution of the teeth

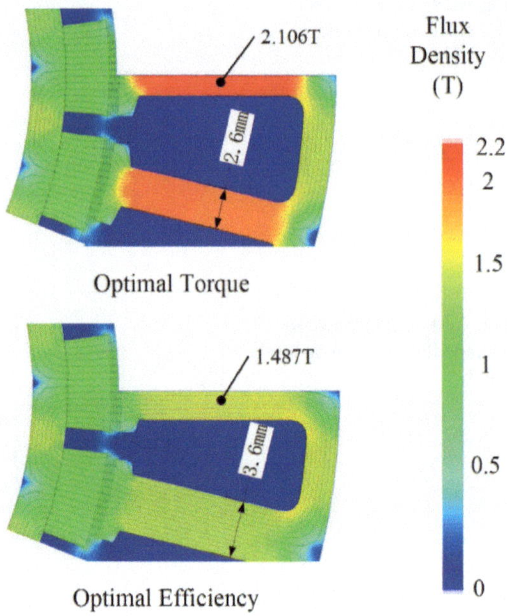

0.24% and –0.08%, indicating the high accuracy of this method. The total losses of the two schemes are 295.8 W and 190.8 W, respectively, with efficiencies of 89.051% and 92.632%. It can be seen that the design method with the best efficiency under given output conditions, compared with the design method with the best moment, reduces losses by 35.50% and improves efficiency by 4.02%.

The magnetic field diagrams of two electromagnetic schemes are shown in Fig. 4.32. Combined with Figs. 4.30 and 4.31, it can be seen that at the optimal torque point, the operating point of the stator core is above the knee point of the BH curve, and the core is close to magnetic saturation. At the optimal efficiency point, the operating point of the stator core is below the knee point of the BH curve.

4.4 Variable Density-Based Multi-Material Topology Optimization for PMSM Rotors

4.4.1 Optimization Method

4.4.1.1 Operating Principle

Figure 4.33 shows the schematic configuration of the magnetic pole. Assume that the rotor design domain is filled with small elements. Assigning different material properties to these elements forms various material distributions within the design

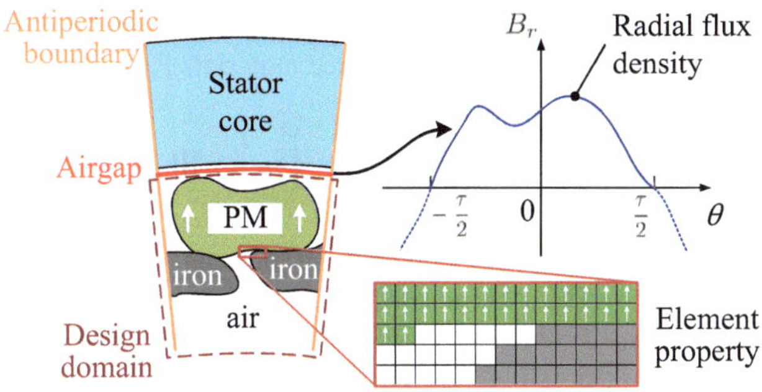

Fig. 4.33 Design concept of the proposed topology optimization method

domain. Material distribution is altered by controlling the material properties within the elements, thereby forming different topological structures. In terms of electromagnetic characteristics, there are three types of materials in the design domain, i.e., permanent magnet, iron and air. Since the radial magnetization pattern is widely employed in electrical machines, this study focuses on the optimization of this magnet pattern. The magnetization direction of the permanent magnet is represented with $\mathbf{e} = (\mathbf{0}, \mathbf{1})$, i.e. the direction of the white arrows in Fig. 4.33.

The magnetic flux density is expressed with magnetic vector potential as

$$\mathbf{B} = \nabla \times \mathbf{A} \tag{4.36}$$

The regions with PM, iron and air are characterized with

$$\mathbf{B} = \begin{cases} \mu_0 \mu_{\text{rec}} \mathbf{H} + \mathbf{B}_{\text{rem}} & \text{, for PM} \\ \mu_0 \mu_{\text{r}} \mathbf{H} & \text{, for iron} \\ \mu_0 \mathbf{H} & \text{, for air} \end{cases} \tag{4.37}$$

where $\mathbf{B}_{\text{rem}}$ is remanence of PM and $\mathbf{B}_{\text{rem}} = \|\mathbf{B}_{\text{rem}}\|\mathbf{e}$, μ_0 is vacuum permeability, μ_{r} is relative permeability, μ_{rec} is the recoil permeability of PM. Since the recoil permeability $\mu_{\text{rec}} \approx 1$ for neodymium magnets and the values of $\|\mathbf{B}_{\text{rem}}\|$ and μ_{r} are variable, the constitutive equation of the three materials can be uniformly expressed as

$$\mathbf{B} = \mu_0 \mu_{\text{r}} \mathbf{H} + \|\mathbf{B}_{\text{rem}}\|\mathbf{e} \tag{4.38}$$

The magnetic field of the rotor is a solenoid field because there is no current passing through it. Thus,

$$\nabla \times \mathbf{H} = 0 \tag{4.39}$$

By calculating the curl on both sides of (4.38) and then substituting (4.36) into it, the corresponding governing equation for three materials is

$$\nabla \times \left(\frac{1}{\mu_0 \mu_r} \nabla \times \mathbf{A} \right) = \nabla \times \left(\frac{1}{\mu_0 \mu_r} \|\mathbf{B}_{\text{rem}}\| \mathbf{e} \right) \tag{4.40}$$

When the relative permeability and remanence of material are assigned with different values, (4.40) represents different materials [23]. By substituting material properties into each element, the electromagnetic field distribution can thus be obtained. Figure 4.33 illustrates a possible distribution of materials and the potential radial magnetic flux density waveform in air gap under the corresponding material distribution state.

Another point to note is that the case with $\mu_r > 1$ & $\|\mathbf{B}_{\text{rem}}\| > 0$ should be avoided since it does not exist in reality for neodymium magnets. For other types of PM materials, such as ferrite type PM, AlNiCo, etc., topology optimization methods are also applicable, albeit with consideration for the non-linearity of magnetic materials [28].

4.4.1.2 Representation of Design Variables

Two density variables are defined to control the material property of each element, i.e., $\alpha_c \in [0, 1]$ and $\beta_c \in [0, 1]$. Density variables are assigned within elements, so numerical artifacts such as checkerboards, zig-zags or wavy boundariess may occur during the optimization process due to mesh dependency. Therefore, Helmholtz filtering method [29] is used for the density variables, i.e.,

$$\alpha_f = R\nabla^2 \alpha_f + \alpha_c, \ \beta_f = R\nabla^2 \beta_f + \beta_c \tag{4.41}$$

where α_f and β_f are filtered density variable, and R depends on the size of elements. As shown in Fig. 4.34, take α for example, the boundary becomes smoother after filtering, but intermediate(grey) variables increase, leading to a fuzzy boundary. However, α and β in every elements should be equal to either 0 or 1 for practical

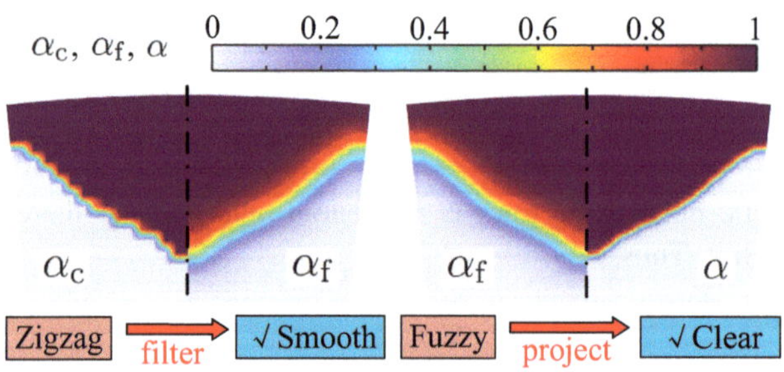

Fig. 4.34 Smooth and clear boundaries after filtering and projection

manufacturing. To reduce the number of intermediate values and make the material boundary clear, the hyperbolic tangent projection [30] is employed to project the density variable to 0 or 1, i.e.

$$\alpha = \frac{\tanh(k_\alpha(\alpha_f - \alpha_0)) + \tanh(k_\alpha \alpha_0)}{\tanh(k_\alpha(1 - \alpha_0)) + \tanh(k_\alpha \alpha_0)}$$
$$\beta = \frac{\tanh(k_\beta(\beta_f - \beta_0)) + \tanh(k_\beta \beta_0)}{\tanh(k_\beta(1 - \beta_0)) + \tanh(k_\beta \beta_0)} \quad (4.42)$$

where k_α and k_β are slope of the projections, and α_0 and β_0 are midpoint of the projections. Then, μ_r and $\|\mathbf{B}_{rem}\|$ can be expressed with α and β [23] as

$$\|\mathbf{B}_{rem}\| = B_{rem}\alpha^m(1 - \beta)^m$$
$$\frac{1}{\mu_r} = 1 + \left[\frac{1}{\mu_{iron}(\mathbf{A})} - 1\right]\beta^n(1 - \alpha)^n, \quad (4.43)$$

where B_{rem} is the remanence of PM. m and n are penalty coefficients of remanence and permeance, respectively. They are employed to weaken the role of intermediate values, and further reduce these values during the optimization process. $\mu_{iron}(\mathbf{A})$ is the relative permeability of the back iron which changes with respect to the magnetic field, i.e.

$$\frac{1}{\mu_{iron}(\mathbf{A})} = \frac{\mu_0 f^{-1}(\|\nabla \times \mathbf{A}\|)}{\|\nabla \times \mathbf{A}\|} \quad (4.44)$$

where f^{-1} is the inverse function of $B = f(H)$. $B = f(H)$ is the relationship between flux density B and magnetic intensity H of a material, which is obtained by magnetization curve test.

The final solution of α and β should be as close as possible to 0 or 1, which is conducive to forming smooth and clear boundariess. Different combinations of α and β can represent three material properties as shown in Table 4.4.

Table 4.4 Relationship between material and density variable

α	β	μ_r	$\|\mathbf{B}_{rem}\|$	Material
0	0	1	0	Air
1	1			
1	0	1	B_{rem}	NdFeB
0	1	μ_{iron}	0	Iron

4.4.1.3 Constraints for the Design

As shown in Table 3.3, when $\alpha = 1$ and $\beta = 1$, the element is filled with air. It does not influence the air gap magnetic field and the torque output. This case is automatically excluded during the optimization. Hence, $\alpha = 1$ represents PM and $\beta = 1$ represents iron. To achieve lightweight of the rotor, the mass of the permanent magnet and the back iron, i.e., the percentage of the areas with $\alpha = 1$ or $\beta = 1$ in the design domain, is constrained. After processing in the previous subsection, the density variable of the elements is either 0 or 1, and there are very few intermediate values. Therefore, the constraints are written as

$$\frac{\int_{\Omega} \alpha ds}{A_{\Omega}} \leq \rho_{\mathrm{PM}}^{*}, \frac{\int_{\Omega} \beta ds}{A_{\Omega}} \leq \rho_{\mathrm{Iron}}^{*} \tag{4.45}$$

where A_{Ω} is the total area of the design domain. ρ_{PM}^{*} and ρ_{Iron}^{*} represent the permissible percentage of PM and iron in design domain, respectively.

The permissible material percentage is a crucial parameter in the topology optimization process, and different material percentages result in distinct distributions of materials. The selection of material percentage is often arbitrary [28].

4.4.1.4 Objective of the Optimization

No matter the magnet-motive force or the back electro-motive force (EMF) is employed to calculate the electromagnetic torque of an electric motor, the average torque output of the motor is proportional to the fundamental air-gap flux density [31]. In other words, the larger the amplitude of the fundamental term of the air-gap flux density, the larger the average electromagnetic torque of the motor. Therefore, optimizing the amplitude of the fundamental term of the air-gap flux density helps to improve the torque output. Since the effect of stator slots on the fundamental magnetic flux density can be corrected by using the Carter coefficient [32], it is ignored here and replaced with a single block of ferromagnetic material as shown in Fig. 4.33. An arc curve in the air gap is used to obtain the waveform of the air-gap magnetic flux density. By performing Fourier decomposition of the air-gap magnetic flux density, the amplitude of the fundamental term of the air-gap flux density can be obtained as

$$B_{r1} = \frac{1}{\tau} \int_{-\tau}^{\tau} B_r(\theta) \cos\left(\frac{\pi \theta}{\tau}\right) d\theta \tag{4.46}$$

where θ is rotor position and τ is a half cycle of radial flux density waveform or a polar distance. $B_r(\theta)$ is the radial flux density at different rotor position. Due to the symmetry of the magnetic poles, only a half period is calculated. Because $\tau = \pi/p$, the objective function is

$$\text{Maximize } B_{r1} = \frac{2p}{\pi} \int_{-\frac{\pi}{2p}}^{\frac{\pi}{2p}} B_r(\theta)\cos(p\theta)d\theta \tag{4.47}$$

The chosen objective function (4.47), instead of directly employing average torque output as objective function, can offer two advantages as follows:

1. *Improved Computational Speed*: Direct computation of the average torque requires the establishment of a simulation model for the individual motor unit, involving transient field simulations or steady-state simulations across multiple rotor positions. This results in a significant increase in computational workload. On the other hand, with the selected objective function (4.47), only one magnetic pole and one steady-state simulation are required in the calculation. This results in a significant reduction in computational workload and an acceleration of the optimization process.
2. *Inherent Structural Symmetry*: The adoption of average torque as the objective function may lead to asymmetrical magnetic pole structures, as demonstrated in the results presented in [27]. In order to avoid asymmetry issues, symmetric constraints are required to be imposed. In contrast, the fundamental wave, explained through trigonometric functions, inherently contains symmetrical information. Therefore, any departure from structural symmetry results in a reduction of the fundamental wave, naturally yielding a symmetric magnetic pole configuration.

4.4.1.5 Design and Optimization Process

The entire design process is illustrated in Fig. 4.35. In order to achieve global optimization, a combination of size optimization and topology optimization has been employed. The specific steps are as follows:

① **Overall size optimization**: Based on requirements of rated operational condition, the size optimization is carried out on the radially magnetized surface-mounted PMSM including both rotor and stator, to obtain optimal structure dimensions.

② **Establishment of FEA model**: Thc topology optimization model is constructed based on the rotor dimensions, the design domain is determined, and meshing on the design domain is performed.

② **Establishment of FEA model**: The topology optimization model is constructed based on the rotor dimensions, the design domain is determined, and meshing on the design domain is performed.

③ **Determination of material proportions**: The material proportions of the permanent magnet and the back iron to the design domain area, i.e. ρ_{PM}^* and ρ_{Iron}^* can be obtained from the ① step. The various combinations of material proportion parameters can be obtained.

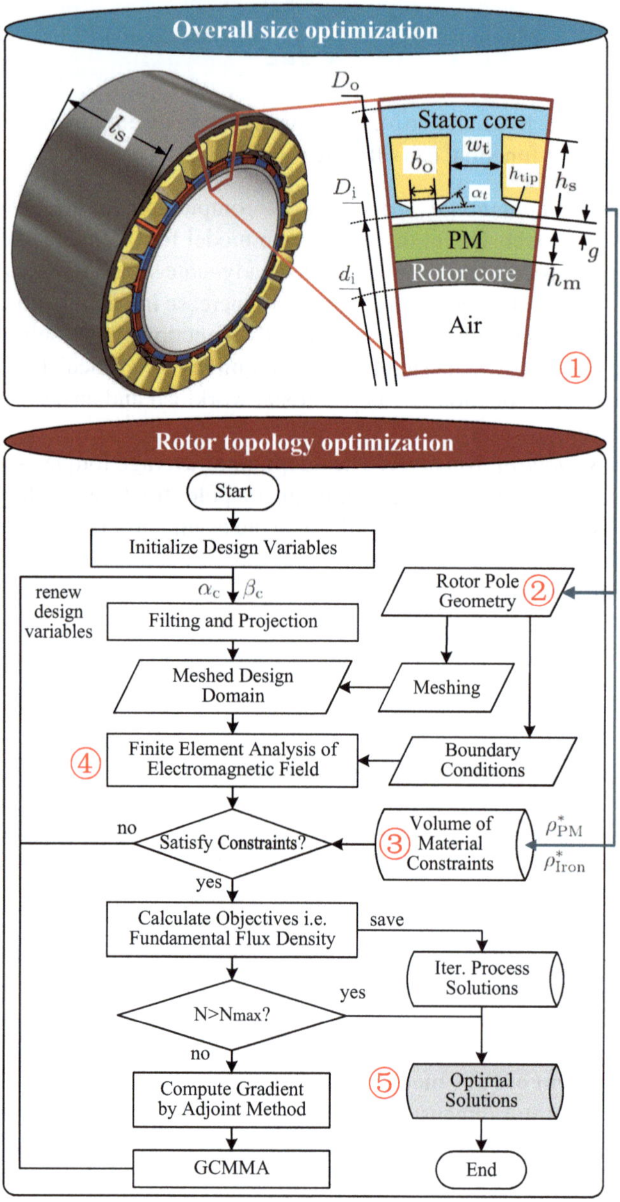

Fig. 4.35 Process flowchart of proposed topology optimization

④ **Topology optimization of pole**: The constraints from the ③ step is employed to perform topology optimization in the design domain from the ② step. Specifically, the initialized design variables are filtered and projected. The FEA computation is conducted to verify the compliance with constraints, and assess whether the objective function increases. The design variables α_c and β_c are updated in every iteration, until the maximum number of iterations.

⑤ **Response surface optimization**: According to results from the ④ step, the response surfaces can be obtained. The optimal material proportions can be determined, and then substituted into the ④ step to obtain the final

In this section, the commercial software COMSOL v6.0 is used for analysis. Globally convergent method of moving asymptotes (GCMMA) [33] along with adjoint method is used as the optimization solver for sensitivity analysis and updating density variables.

4.4.2 *Example of Optimization*

4.4.2.1 Problem Description and Establishment of FEA Model

An aerospace motor of air compressor pump under rated speed of 1600 r/min and power of 335 W is design and optimized for high torque density by conventional size optimization [34]. The lamination B35A250 and the rare earth NdFeB N40UH are utilized for the stator cores and the magnets, respectively. The rotor is made of solid stainless steel 20Cr13 for ease of manufacturing. To obtain the non-linear material properties of 20Cr13, a ring-shaped specimen was processed (as shown in Fig. 4.36a), and its BH curve was then obtained with the standard IEC 60404-4 [35], as shown in Fig. 4.36b where $\beta = 1$. Finally, the major parameters of the optimized electric motor with original configuration as shown in Fig. 4.35 are listed in Table 4.5.

Extracting a pole pitch of the motor, a FEA model is established as shown in Fig. 4.37a. The inner diameter d_s of the design domain is arbitrarily set to 48 mm, which is smaller than the optimized value d_i. A curved line inside the air gap rep-

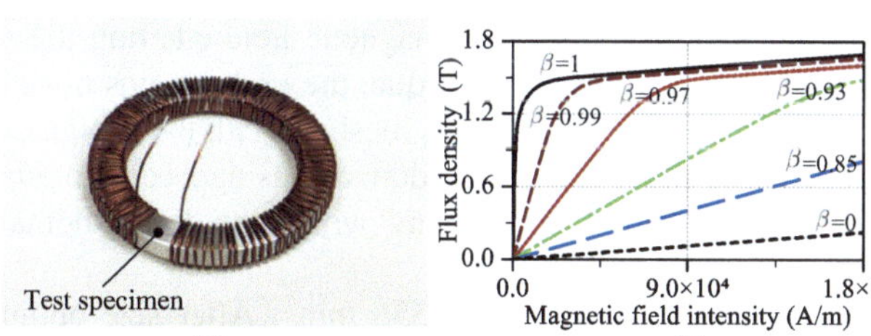

Fig. 4.36 The BH test specimen and BH curve for the rotor iron material: **a** Test specimen; **b** BH curve and that with different density variables

Table 4.5 Parameters of the motor to be optimize

Notation	Geometric parameter	Value	Unit
Q	Slot number	27	
p	pole number	30	
D_o	Stator diameter	82	mm
D_i	Stator bore	61	mm
h_s	Slot depth	9	mm
w_t	Tooth width	2.7	mm
h_{tip}	Tooth tip depth	0.5	mm
α_t	Tooth tip angle	25	deg
b_o	Slot opening	1.1	mm
l_s	Stator length	36	mm
g	Air gap	0.4	mm
d_o	Rotor diameter	60.2	mm
d_i	Rotor bore	54	mm
h_m	Height of PM	1.8	mm
J_{rms}	RMS current density	2.8	A/mm^2
k_f	Copper slot fill	0.2824	

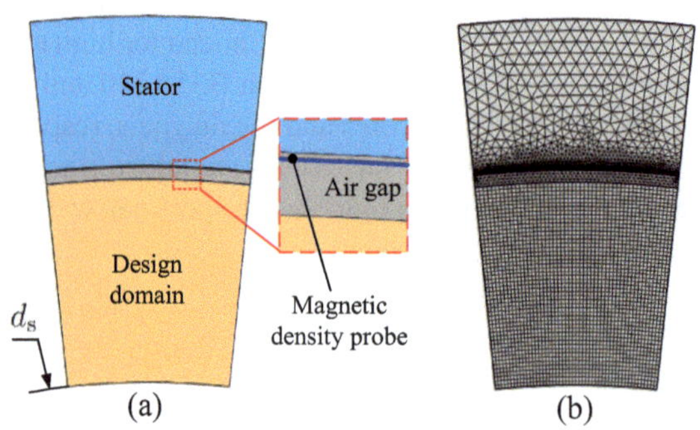

Fig. 4.37 Finite element model: **a** Geometric layout; **b** Mesh of the model

resents a magnetic probe used to calculate the fundamental magnetic flux i.e. topology optimization objectives. As only the magnetic field entering the stator region can interact to produce electromagnetic torque, the probe is positioned as close as possible to the stator bore. The geometry is meshed, with the design domain using a structured grid, and the stator and air gap domains using free triangular meshes as shown in Fig. 4.37b. There are 2500 elements within the design domain and 4964 elements in the entire computational model.

The design domain has an area of 34.558 mm^2. After size optimization, the cross-sectional areas of the rotor magnetic steel and back iron are 11.008 mm^2 and 7.528 mm^2, respectively, representing percentages of 0.3185 and 0.2178 relative to the design domain. The optimization objective of this study is to further increase torque density by using fewer materials of PM and back iron. Therefore, the material percentage of PM and iron should be even smaller, i.e. $\rho^*_{\rm PM} \in [0, 0.3185]$ and $\rho^*_{\rm Iron} \in [0, 0.2178]$.

4.4.2.2 Topological Optimization of Rotor Magnetic Poles

The values for the projection parameters in (4.42) are set to $k_\alpha = k_\beta = 10$ and $\alpha_0 = \beta_0 = 0.5$. The values for the penalty coefficients in (4.43) are set to $m = 6$ and $n = 2$. Under this configuration, the dependency of magnetic flux density on magnetic field intensity with different β in the region where $\alpha = 0$ is shown in Fig. 4.36b. And the initial values of density variables are arbitrarily set to $\alpha_{\rm c} = \beta_{\rm c} = 0.5$ for all elements in design domain, which has no impact on the optimization results.

As an example with $\rho^*_{\rm PM} = 0.1911$ and $\rho^*_{\rm Iron} = 0.1089$, the optimization process is shown in Fig. 4.38. In the initial state, α and β are evenly distributed according to initial value. Then, α and β decrease simultaneously to satisfy the constraints. From the 2nd step, α near by the top of the rotor pole begins to grow, which results in the PM region. From the 4th step, β on both sides of the pole gradually increases, which results in the iron region. As the number of iterations increases, the intermediate variables decrease gradually, and the boundariess become clearer. After 25 iterations, the clear boundary is obtained. As shown in Fig. 4.39, the value of the objective function hardly changes thereafter, and the percentage of the two materials meet the constraints.

The entire process was iterated 50 times, taking 269 s, with an average of 5 s per iteration. Comparing with [27], which utilizes the average torque as the objective function, the computation speed is improved enormously. This because that the value

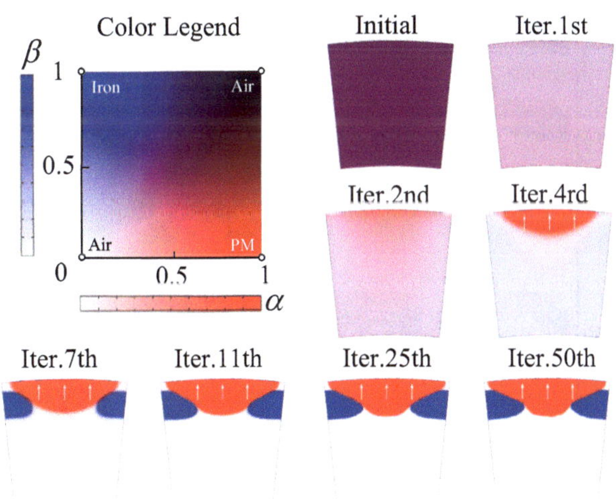

Fig. 4.38 Iteration optimization processing and typical solutions

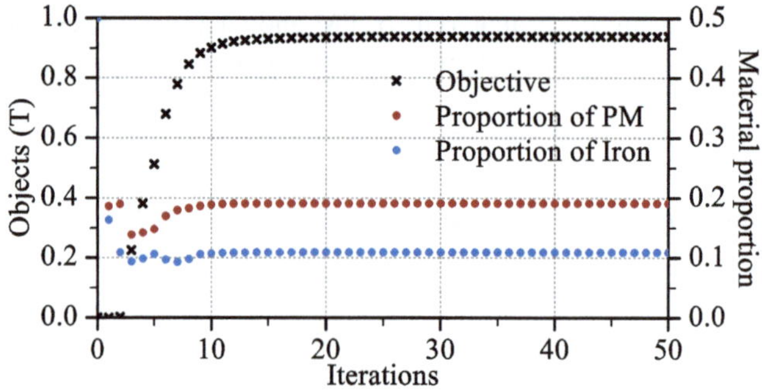

Fig. 4.39 Objective function changes with each iterations

of average torque requires the calculation of the torque at N positions while the proposed method that employs the air-gap fundamental flux density as the objective function only requires the calculation of one tooth to evaluate the average torque. Thus, it can increase the computing speed significantly.

4.4.2.3 Response Surface Optimization

To determine the optimal permissible percentages of materials, different percentages were selected for the topology optimization of the rotor poles, resulting in 20 configurations as shown in Fig. 4.40. It is evident that the permissible percentages of materials has a significant impact on the optimization results. By substituting these

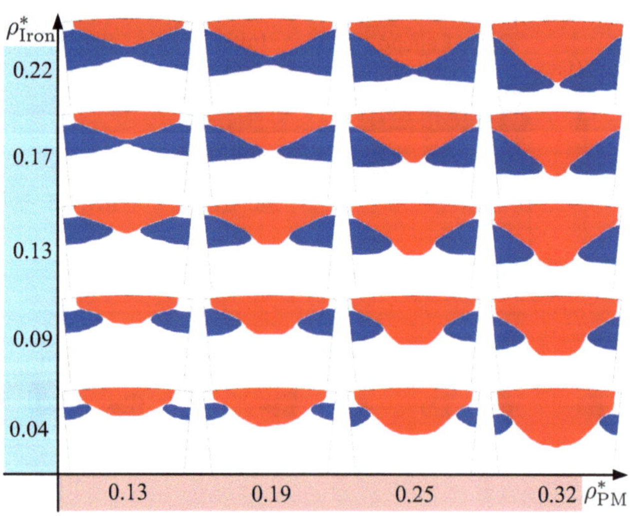

Fig. 4.40 Optimization results with different permissible portion of materials

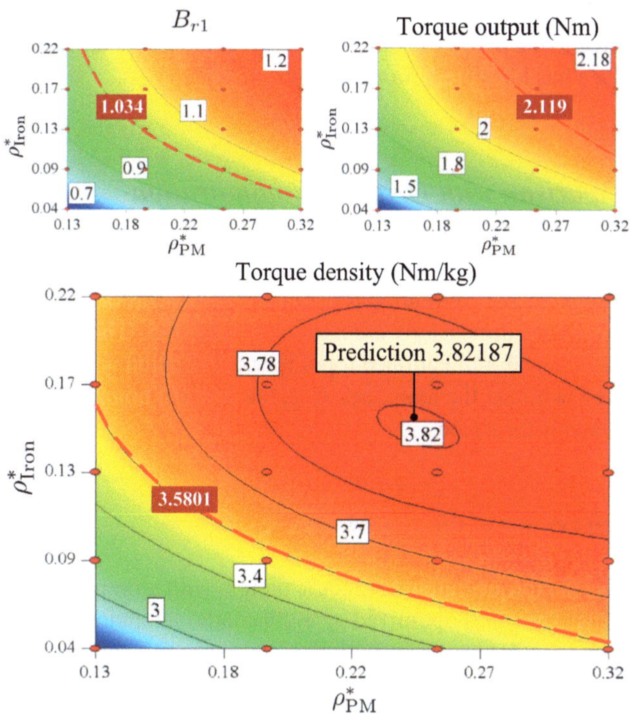

Fig. 4.41 Response surfaces of flux density, torque and torque density

20 configurations for the original rotor pole, the fundamental radial air-gap flux density, torque output and effective torque density of the motor were calculated, forming the response surfaces as shown in Fig. 4.41. These response surfaces illustrate the influence of different percentage of materials on the fundamental radial air-gap flux density, torque output and effective torque density of the motor. The maximum values of permissible percentages in the response surfaces come from the usage of PM and back iron obtained through the motor's global optimization with the original rotor pole configuration, so as the performance level highlighted by the dashed line in the figure.

From Fig. 4.41, it can be observed that the usage of PM and soft magnetic material has a similar impact on the fundamental radial air-gap flux density and torque output of motor. As the material usage increases, the improvement in both fundamental flux density and torque output gradually slows down. Therefore, it is not wise to increase the output torque by simply increasing the material usage. In addition, the maximum value of torque density is located at $\rho^*_{\mathrm{PM}} = 0.2465$ and $\rho^*_{\mathrm{Iron}} = 0.1537$ in the range of response surfaces. Hence, utilizing the usage of permanent magnets and back iron obtained through global optimization to narrow down the range of permissible percentages of materials is reasonable.

The torque density for the optimal validation is 3.81025 Nm/kg, with an error of –0.3% compared to the predicted value of 3.8219 Nm/kg, indicating that the obtained response surface has high prediction accuracy.

4.4.3 Comparison with Original Configuration

The two optimized magnet pole patterns are compared as shown in Fig. 4.42. The left one is optimized by conventional parameter optimization method with radial magnetized poles. Its dimensions are shown in Table 4.5. The right one is reconstructed based on the aforementioned topology optimization where $\rho_{\mathrm{PM}}^{*} = 0.28$ and $\rho_{\mathrm{Iron}}^{*} = 0.14$. To facilitate the manufacturing, the internal boundary of the back iron is fitted using a circular shape.

4.4.3.1 Comparisons of the Two Motors

Table 4.6 shows the torque output of electric motors with aforementioned two different pole patterns. The poles optimized by multi-material topology method costs less PM material by 9.45% than those optimized by conventional size design method, and back iron by 33.53%. Hence, the weight of rotor decreases by 19.42%. However the average torque output is increased by 1.23%, ultimately resulting in the increase of torque density by 6.55%. As shown in Fig. 4.43, the torque density of the optimized design is larger than that of the conventional one. The improvement of the PM utilization ratio makes the motor more cost-efficiency. And the decrease of the moment of inertia can improve the dynamic characteristics of the motor. In addition, the cogging torque is reduced, and the efficiency is improved, taking into account the stator iron loss, stator copper loss, magnetic steel eddy loss, and rotor back iron eddy loss.

For single or small batch production, the wire electrical discharge machining is usually utilized for fabrication. The usage of PM material and the corresponding cost can be reduced by using the proposed optimization method, and the complicated magnet shape may increase the cost a bit. For mass production, the mould can be employed for the manufacturing to reduce the total cost.

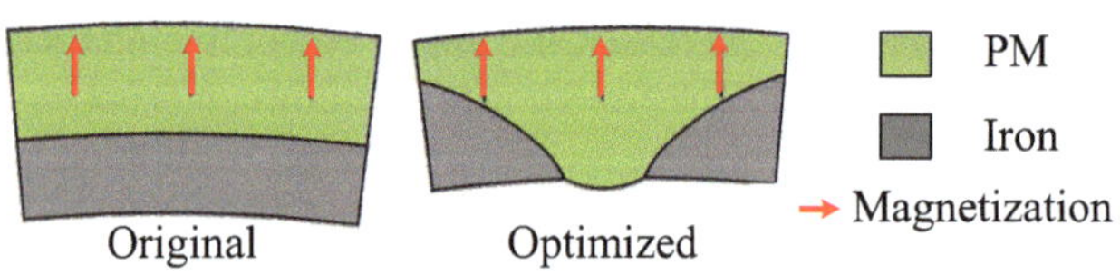

Fig. 4.42 Two patterns optimized by conventional and proposed method

Table 4.6 Relationship between multiple material properties

Item	Unit	Original	Optimized	Improvement (%)
Magnet weight	Kg	0.08917	0.08074	–9.45
Rotor iron weight	kg	0.06301	0.04188	–33.53
Total weight (active)	kg	0.59188	0.56232	–4.99
Average torque	N·m	2.119	2.145	+1.23
Cogging torque	mN·m	12.528	10.769	–14.04
Efficiency	%	91.022	91.072	+0.05
Torque density	Nm/kg	3.5801	3.8145	+6.55

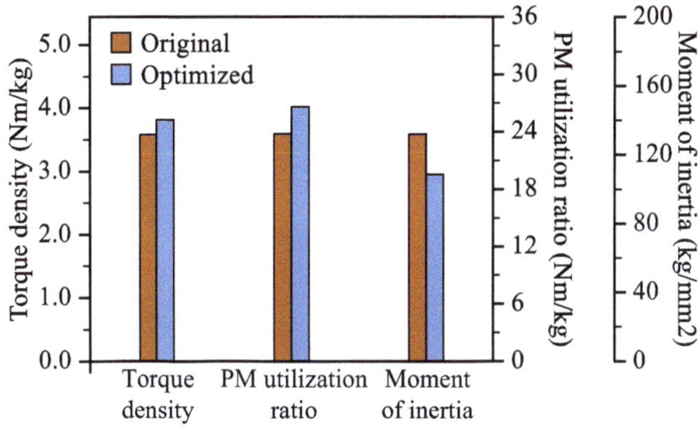

Fig. 4.43 Comparisons of torque density, PM utilization and Moment of inertia

4.4.3.2 Magnetic Field and Torque Output

The waveforms of the air gap magnetic flux density for the two kinds of pole patterns are shown in Fig. 4.44a. The magnetic flux density waveforms are similar, but the optimized one in this study has slightly higher peaks than the conventional one. The fast Fourier transform is performed on the two waveforms, and the results are shown in Fig. 4.44b. It can be observed that the fundamental component, or the 1st harmonic of the optimized pole is larger than that of the conventional one, while the 3rd, the 5th and the 7th harmonics are smaller. The harmonics distribution can not only achieve high torque density but also reduce torque ripple and iron loss [36–38].

Figure 4.45 shows the magnetic field simulation of the two motors. It can be seen that the magnetic flux density in the back iron is evenly distributed, while the magnetic flux density distribution for the conventional design is uneven, i.e., some portions are saturated and other portions have little magnetic flux passing through. The material utilization rate is higher than that of the conventional one, which explains why the torque is increased while the rotor mass is decreased.

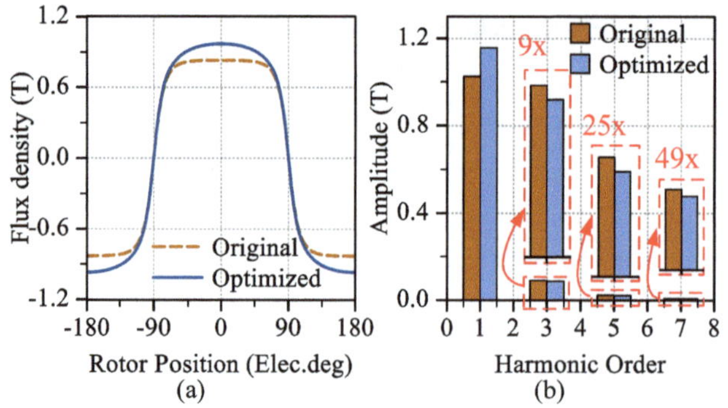

Fig. 4.44 Radial flux density waveforms in air gap. **a** shape of the waveforms; **b** FFT of the waveforms

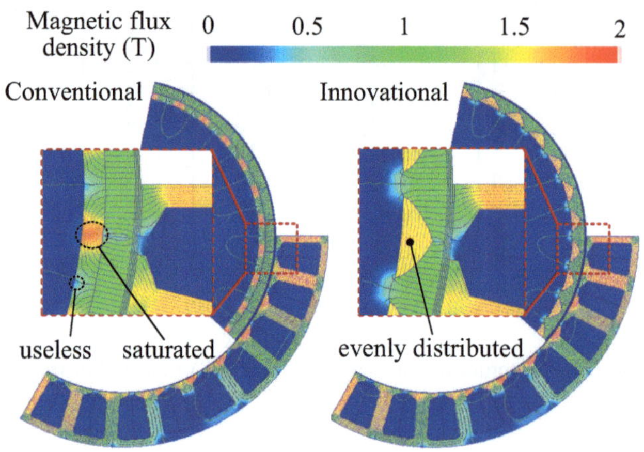

Fig. 4.45 Electromagnetic field of motors with different pole patterns

4.4.4 *Research Prototype and Experiments*

4.4.4.1 Development of Research Prototype

One research prototype of the electric machine with the optimized configuration is developed as shown in Fig. 4.46. The PM pole and the outer profile of the rotor core are manufactured by wire electrical discharge machining, and the surface of the PM is nickel plated. The PM poles are glued to the rotor core, and the strength and stiffness of the rotor are verified to be more than enough by FEA. As shown in Fig. 4.47, the Y connection is adopted and the neutral point is led out to facilitate the measurement of back EMF. Two PT-1000 temperature sensors are inserted between the windings to monitor the winding temperature. In order to ensure the coaxiality between the stator and the rotor, the housing rabbet is turning processed based on

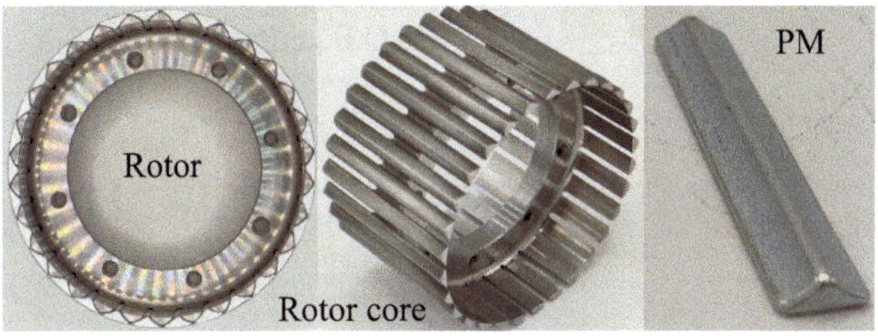

Fig. 4.46 Rotor with optimized poles and rotor core

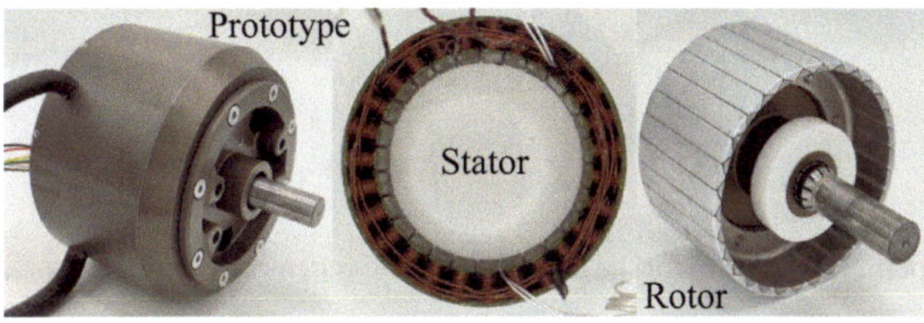

Fig. 4.47 Research prototype and major components of the motor

the stator bore after installed into by the stator. Besides, the axial clearance of the bearings is adjusted to ensure smooth and stable rotation of the rotor. To avoid axial misalignment of the rotor, the axial length of the rotor has been extended by 1 mm. The housing and the end cap are made of aluminum alloy 7075. Although the shaft and the rotor use the same magnetic material, non-magnetic aluminum alloy is used to connect them, to achieve magnetic circuit isolation and avoid the influence of the shaft on the rotor magnetic circuit. This stand structure ensures consistency between simulation and experiment.

4.4.4.2 Experiments on Torque Output

Experimental test rig for the torque output measurement is shown in Fig. 4.48. The research prototype is fixed on the three jaw chuck of the indexing plate so that the stator can rotate to any specific angle. The output shaft is fixed on the rotor through a torque transducer (ATI F/T Mini45), so that the torque output of the electric motor can be measured and collected by the computer through DAQ equipment (NI-DAO-USB 6002). A resolver is installed on the output shaft to read the relative position between the stator and the rotor so as to realize the current loop control with $I_d = 0$ by the servo drive. The electromagnetic torque is generated when I_q or peak current is assigned with any value above 0. The electromagnetic torque at specific rotation angle is also obtained by rotating the stator to corresponding angle. In addition, the temperature sensors inside the winding are connected to a monitor to collect the temperature of the windings.

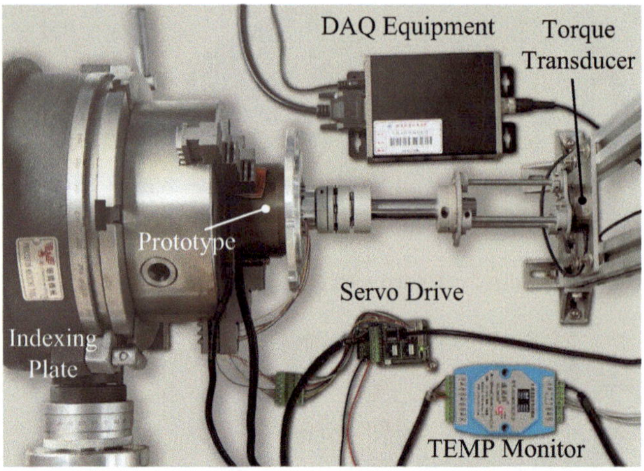

Fig. 4.48 Experimental test rig for the torque output measurement

As shown in Fig. 4.49, the torque output obtained through experiments and the simulation results are compared. The average torque output of the experimental and simulation results is consistent. Figure 4.49 also presents the torque within one electrical cycle when $I_q = 10$ A. The torque ripple of experimental result is slightly larger than that of simulation, which may be caused by the manufacturing and assembly error [39] of the research prototype.

4.4.4.3 Experiments on Back Electromotive Force

The experiments on the back EMF of the research prototype is also conducted as shown in Fig. 4.50. The prototype is dragged by another motor which is controlled by a servo drive. The back-EMF waveform of the motor is recorded by an oscilloscope (RIGOL DH04204) when the electric motor is rotated at a specific speed. Any phase back EMF can be directly collected, since the neutral point is led out.

The comparison of experimental result and simulation is shown in Fig. 4.51. When the shaft speed is 1600 r/min, the back EMF from the experiment is consistent with that of simulation. The other measurement results are slightly larger than the simulation results, possibly due to mismatched driver parameters, resulting in an increase in peak to peak values caused by speed fluctuations at low speeds.

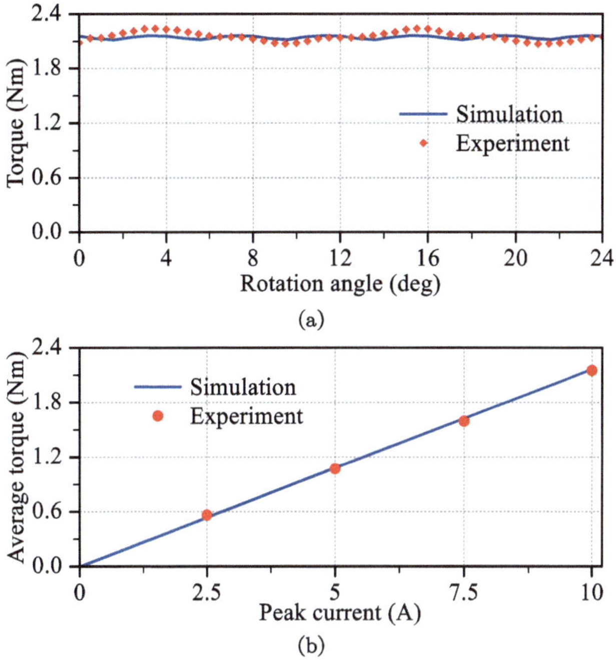

Fig. 4.49 Torque comparison between experimental and simulation. **a** Torque profile. **b** Average torque

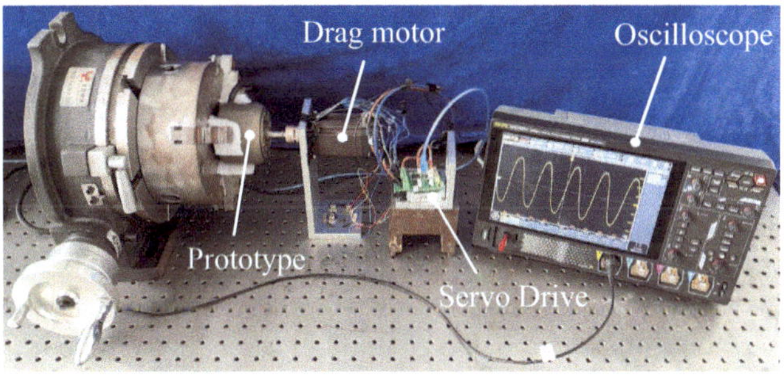

Fig. 4.50 Experimental test rig for the EMF measurement

4.5 Conclusion

Firstly, this chapter compares the impact of various density interpolation functions on the results of multi-material topology optimization of motors. It was found that the results of the permeability interpolation based on magnetic flux density were unstable. In response to this phenomenon, two solutions were compared and both the averaging method and the continuation method were effective in addressing the

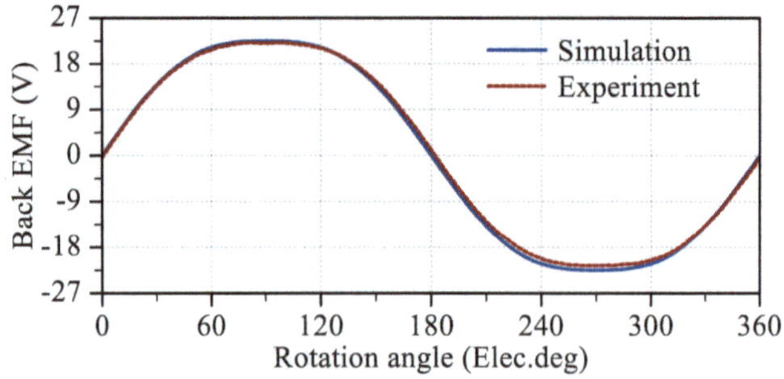

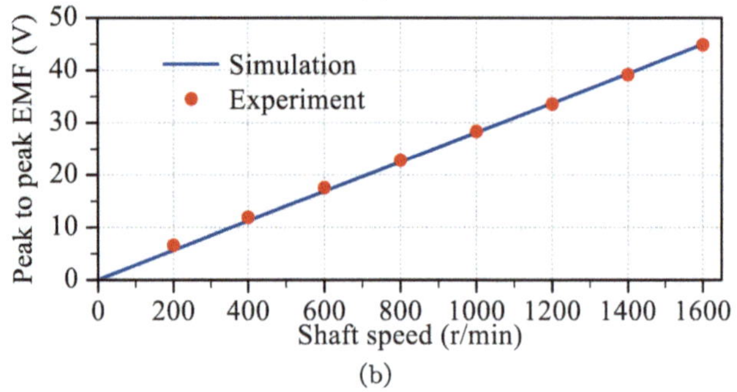

Fig. 4.51 Back EMF comparison between experiment and simulation. **a** Back EMF profile. **b** Peak-to-peak EMF

computational instability. However, due to the significant impact of intermediate variables on the results, the practical utility of the obtained results was limited. To address computational instability, the key lies in selecting an appropriate interpolation function. The chapter explains the intrinsic connection between density interpolation functions and electromagnetic field calculation methods. Specifically, the choice of interpolation function should be related to the electromagnetic field calculation method. The specific rule is that the dependent variable in field calculation should be in a product relationship with the interpolation function, rather than being placed in the denominator of the dependent variable. Otherwise, small variations will have a significant impact on the results, leading to computational instability. This hypothesis was validated by adopting the permeability interpolation based on magnetic scalar potential electromagnetic field calculation, which also yielded excellent optimization results.

Secondly, in this chapter, a multi-material topology optimization method based on variable density representation is proposed for the design optimization of surface-mounted permanent-magnet motors. The topological configurations of rotor magnetic poles are obtained by optimizing material distribution, rather than fixing the magnetic pole shape from the beginning. This proposed optimization method helps to achieve higher torque density of surface-mounted PMSM, for which the torque output are proportional to the fundamental magnetic flux density of the air gap. The main contribution lies in improving the optimization speed by improving the objective function, and achieving clear instead of arbitrary volume constraints by combining size optimization and topology optimization. The major works are concluded as follows:

1. The fundamental concept and procedures of the optimization method including the design variables, the constraints and the optimization objective are presented. The advantages of the optimization method are introduced.
2. The proposed optimization method is implemented into the design of permanent-magnet motor with radial magnetization poles, and its result is compared with that of the conventional optimization method. The comparison shows that the machine design by the proposed method can improve the magnetic flux density and the torque output, and reduce the rotor mass simultaneously, which in turn indicates that it helps to achieve high torque density.
3. One research prototype of the electric motor with the optimized structure has been developed, and experiments are conducted on the torque output and the back EMF. The experimental results are compared with the numerical simulation, and the comparison shows the consistency between. Thus, the proposed design method is validated.

Meanwhile, this chapter also proposes a motor design method with fixed output operating conditions, which simplifies the calculation process and can accurately obtain the optimal efficiency solution under rated operating conditions. Sensitivity analysis was conducted on the stator tooth width of existing electromagnetic schemes, and two optimal solutions were obtained, namely the torque optimal solution and the solutions, it was found that the stator teeth of the torque optimal solution worked above the knee point of its BH curve, and the stator teeth were close to saturation; The stator teeth of the optimal efficiency solution work below the knee point of its BH curve. Under rated operating conditions, the performance of the efficiency optimal solution is better than that of the torque optimal solution.

References

1. X. Ba, X. Sun, Z. Gong, Y. Guo, C. Zhang, and J. Zhu, "A generalized per-phase equivalent circuit model of the pmsm with predictable core loss," IEEE/ASME Trans. Mechatronics, vol. 28, no. 3, pp. 1512–1521, 2023.
2. W. Geng, J. Wang, L. Li, and J. Guo, "Design and multiobjective optimization of a new flux-concentrating rotor combining halbach pm array and spoke-type ipm machine," IEEE/ASME Trans. Mechatronics, vol. 28, no. 1, pp. 257–266, 2023.
3. B. A. Adu-Gyamfi and C. Good, "Electric aviation: A review of concepts and enabling technologies," Transportation Engineering, vol. 9, Sep. 2022.
4. P. Wheeler, "Technology for the more and all electric aircraft of the future," 2016 IEEE International Conference on Automatica (ICA- ACCA), Oct. 2016, pp. 1–5.
5. E. Sayed et al., "Review of Electric Machines in More-/Hybrid-/Turbo-Electric Aircraft," in IEEE Transactions on Transportation Electrification, vol. 7, no. 4, pp. 2976–3005, 2021.

6. J. Pyrhonen, T. Jokinen, and V. Hrabovcova, "Design of Rotating Electrical Machines," John Wiley & Sons, Sep. 2013.
7. Z. Zhu, Z. Xia, and D. Howe, "Comparison of Halbach magnetized brushless machines based on discrete magnet segments or a single ring magnet," IEEE Trans. Magn., vol. 38, no. 5, pp. 2997–2999, 2002.
8. P. Salminen, "Fractional slot permanent magnet synchronous motors for low speed applications," Acta Universitatis Lappeenrantaensis, vol. 6, 2004.
9. T. Heikkilae, "Permanent magnet synchronous motor for industrial inverter applications-analysis and design," Acta Universitatis Lappeenrantaensis, vol. 4, 2002.
10. Y. Yang, X. Wang, R. Zhang, T. Ding, and R. Tang, "The optimization of pole arc coefficient to reduce cogging torque in surface-mounted permanent magnet motors," IEEE Trans. Magn., vol. 42, no. 4, pp. 1135–1138, 2006.
11. P. Zheng, J. Zhao, J. Han, J. Wang, Z. Yao, and R. Liu, "Optimization of the Magnetic Pole Shape of a Permanent-Magnet Synchronous Motor," IEEE Trans. Magn., vol. 43, no. 6, pp. 2531–2533, 2007.
12. Y. Zeng, M. Cheng, G. Liu, and W. Zhao, "Effects of Magnet Shape on Torque Capability of Surface-Mounted Permanent Magnet Machine for Servo Applications," IEEE Trans. Ind. Electron., vol. 67, no. 4, pp. 2977–2990, Apr. 2020.
13. Y. Shen and Z. Q. Zhu, "Investigation of Permanent Magnet Brushless Machines Having Unequal-Magnet Height Pole," IEEE Trans. Magn., vol. 48, no. 12, pp. 4815–4830, 2012.
14. Z. Q. Zhu, K. Wang, and G. Ombach, "Optimal magnet shaping with third order harmonic for maximum torque in brushless AC machines," Bristol, UK, Jan. 2012, pp. 81–81.
15. K. Wang, Z. Q. Zhu, and G. Ombach, "Torque Improvement of Five- Phase Surface-Mounted Permanent Magnet Machine Using Third-Order Harmonic," IEEE Trans. Energy Convers., vol. 29, no. 3, pp. 735–747, 2014.
16. A. H. Isfahani, S. Vaez-Zadeh, and M. A. Rahman, "Using Modular Poles for Shape Optimization of Flux Density Distribution in Permanent- Magnet Machines," IEEE Trans. Magn., vol. 44, no. 8, pp. 2009–2015, 2008.
17. M. Ashabani and Y. A.-R. I. Mohamed, "Multiobjective Shape Opti-mization of Segmented Pole Permanent-Magnet Synchronous Machines With Improved Torque Characteristics," IEEE Trans. Magn., vol. 47, no. 4, pp. 795–804, 2011.
18. P. Xiang, L. Yan, Y. Guo, X. He, C. Gerada, and I.-M. Chen, "A Concentrated-Flux-Type PM Machine With Irregular Magnets and Iron Poles," IEEE/ASME Trans. Mechatronics, pp. 1–12, 2023.
19. R. T. Haftka and Z. Gürdal, "Elements of structural optimization," Springer Science & Business Media, vol. 11, 2012.
20. J. Gao, M. Xiao, Y. Zhang, and L. Gao, "A Comprehensive Review of Isogeometric Topology Optimization: Methods, Applications and Prospects," Chinese Journal of Mechanical Engineering, vol. 33, no. 1, Nov. 2020.
21. D. Dyck and D. Lowther, "Automated design of magnetic devices by optimizing material distribution," IEEE Trans. Magn., vol. 32, no. 3, pp. 1188–1193, 1996.
22. J. S. Choi, J. Yoo, S. Nishiwaki, and K. Izui, "Optimization of magnetization directions in a 3-d magnetic structure," IEEE Trans. Magn., vol. 46, no. 6, pp. 1603–1606, 2010.
23. J. Lee, M. Yoon, T. Nomura, and E. M. Dede, "Topology optimization for design of segmented permanent magnet arrays with ferromagnetic materials," Journal of Magnetism and Magnetic Materials, vol. 449, pp. 571–581, 2018.
24. T. Jung, J. Lee, and J. Lee, "Design and fabrication of magnetic system using multi-material topology optimization," IEEE Access, vol. 9, pp. 8649–8658, 2021.
25. T. Gauthey, P. Gangl, and M. H. Hassan, "Multi-material topology optimization with continuous magnetization direction for motors design," in 2022 International Conference on Electrical Machines (ICEM), 2022, pp. 483–489.
26. C. Lee and I. G. Jang, "Multi-material topology optimization for the PMSMs under the consideration of the MTPA control," Structural and Multidisciplinary Optimization, vol. 65, no. 9, Sep. 2022.

27. T. Cherriere, S. Hlioui, L. Laurent, F. Louf, H. B. Ahmed, and M. Gabsi, "Topology optimization of asymmetric pmsm rotor," 2022 International Conference on Electrical Machines (ICEM), 2022, pp. 469–475.
28. T. Cherriere, T. Vancorsellis, S. Hlioui, L. Laurent, F. Louf, H. B. Ahmed, and M. Gabsi, "A multi-material topology optimization considering the pm nonlinearity," IEEE Trans. Magn., vol. 59, no. 5, pp. 1–9, 2023.
29. B. S. Lazarov and O. Sigmund, "Filters in topology optimization based on helmholtz-type differential equations," International Journal for Numerical Methods in Engineering, vol. 86, no. 6, pp. 765–781, 2011.
30. F. Wang, B. S. Lazarov, and O. Sigmund, "On projection methods, convergence and robust formulations in topology optimization," Structural and Multidisciplinary Optimization, vol. 43, no. 6, pp. 767–784, 2011.
31. A. E. Fitzgerald, C. Kingsley, and S. D. Umans, "Electric machinery," 2003.
32. F. Carter, "Air-gap induction," Electrical World and Engineer, vol. 38, no. 22, pp. 884–888, 1901.
33. K. Svanberg, "A Class of Globally Convergent Optimization Methods Based on Conservative Convex Separable Approximations," SIAM Journal on Optimization, vol. 12, no. 2, pp. 555–573, 2002.
34. X. He, L. Yan, P. Xiang, N. Du, X. Liu, and J. Liu, "Multi-objective optimization method for surface-mounted PMSM with fixed torque output," 2023 26th International Conference on Electrical Machines and Systems (ICEMS), 2023, pp. 4904–4908.
35. F. Fiorillo, "Measurements of magnetic materials," Metrologia, vol. 47, no. 2, Mar. 2010.
36. Z. Dong, L. Yan, H. Su, and I.-M. Chen, "Design and Analysis of a Novel Electromagnetic-Propulsion Rotary Machine With External Blending-Shaped Tessellation Magnet Pattern," IEEE Trans. Ind. Electron., pp. 1–10, 2023.
37. P. Xiang, L. Yan, H. Xiao, Xinghua He, and Nannan Du, "Development of a Novel Radial-Flux Machine With Enhanced Torque Profile Employing Quasi-Cylindrical PM Pattern," IEEE Trans. Energy Convers., pp. 1–12, 2023.
38. Z. Xia, Z. Zhu, and D. Howe, "Analytical magnetic field analysis of Halbach magnetized permanent-magnet machines," IEEE Trans. Magn., vol. 40, no. 4, pp. 1864–1872, 2004.
39. J. Ou, Y. Liu, R. Qu, and M. Doppelbauer, "Experimental and theoretical research on cogging torque of PM synchronous motors considering manufacturing tolerances," IEEE Trans. Ind. Electron., vol. 65, no. 5, pp. 3772–3783, 2018.
40. D. G. Dorrell, "A Review of the Methods for Improving the Efficiency of Drive Motors to Meet IE4 Efficiency Standards," Journal of Power Electronics, vol. 14, no. 5, pp. 842–851, 2014.
41. S. N. Vukosavic and E. Levi, "A method for transient torque response improvement in optimum efficiency induction motor drives," IEEE Transactions on Energy Conversion, vol. 18, no. 4, pp. 484–493, 2003.
42. S. Bu, L. Yan, X. Gao, P. Zhao, and C. K. Lim, "Vision-guided manipulator operating system based on CSRT algorithm," International Journal of Hydromechatronics, vol. 5, no. 3, pp. 260–274, 2022.
43. Z. Dong, L. Yan, H. Su, and I.-M. Chen, "Design and Analysis of a Novel Electromagnetic-Propulsion Rotary Machine With External Blending-Shaped Tessellation Magnet Pattern," IEEE Transactions on Industrial Electronics, pp. 1–10, 2023.
44. A. Schoppa and P. Delarbre, "Soft Magnetic Powder Composites and Potential Applications in Modern Electric Machines and Devices," IEEE Transactions on Magnetics, vol. 50, no. 4, pp. 1–4, Apr. 2014. "The effects of grain size on the core loss and permeability of motor lamination steel," IEEE Transactions on Magnetics, vol. 22, no. 2, pp. 101–106, Mar. 1986.
45. S. W. Cheong, E. J. Hilinski, and A. D. Rollett, "Grain growth in a low-loss cold-rolled motor-lamination steel," Metall Mater Trans A, vol. 34, no. 6, pp. 1321–1327, 2003.
46. P. Xiang, L. Yan, Y. Guo, X. He, C. Gerada, and I.-M. Chen, "A Concentrated-Flux-Type PM Machine With Irregular Magnets and Iron Poles," IEEE/ASME Transactions on Mechatronics, pp. 1–12, 2023.

47. V. Sarac, "Design Optimization of Induction Motor for Efficiency and Reliability Improvement," 30th International Conference on Organization and Technology of Maintenance (OTO 2021), H. Glavas, M. Hadzima-Nyarko, M. Karakasic, N. Ademovic, and S. Avdakovic, Eds., in Lecture Notes in Networks and Systems. Cham: Springer International Publishing, 2022, pp. 35–49.
48. A. Dietz, A. O. Di Tommaso, F. Marignetti, R. Miceli, and C. Nevoloso, "Enhanced Flexible Algorithm for the Optimization of Slot Filling Factors in Electrical Machines," Energies, vol. 13, no. 5, Jan. 2020.
49. H. Hayashi, K. Nakamura, A. Chiba, T. Fukao, K. Tungpimolrut, and D. G. Dorrell, "Efficiency Improvements of Switched Reluctance Motors With High-Quality Iron Steel and Enhanced Conductor Slot Fill," IEEE Transactions on Energy Conversion, vol. 24, no. 4, pp. 819–825, 2009.
50. M. Abdalmagid, E. Sayed, M. H. Bakr, and A. Emadi, "Geometry and Topology Optimization of Switched Reluctance Machines: A Review," IEEE Access, vol. 10, pp. 5141–5170, 2022.
51. P. Xiang et al., "Design and Optimization of BLDC Machine for Bidirectional Impeller Pump," 2021 IEEE 16th Conference on Industrial Electronics and Applications (ICIEA), Aug. 2021, pp. 2146–2150.
52. X. He, L. Yan, Xiaoshan Gao, Pengjie Xiang, and Chris Gerada, "Comparative Study of Air-Gap Field Distribution Characteristics Among Different Pole Array in SPM," presented at the International Conference of Fluid Power and Mechatronic Control Engineering (ICFPMCE 2022), Atlantis Press, Dec. 2022, pp. 202–213.
53. P. Xiang, L. Yan, H. Xiao, Xinghua He, and Nannan Du, "Development of a Novel Radial-Flux Machine With Enhanced Torque Profile Employing Quasi-Cylindrical PM Pattern," IEEE Transactions on Energy Conversion, pp. 1–12, 2023.
54. Y. Li, L. Yan, P. Xiang, A. Aboelhassan, and C. Gerada, "Design optimization of double-sided PMLSM with unequal width Halbach magnet array," CSAA/IET International Conference on Aircraft Utility Systems (AUS 2022), Aug. 2022, pp. 740–744.
55. X. Wei, L. Yan, Z. Jin, Y. Zhou, and P. Xiang, "SPMSM Deadbeat Predictive Control with HF Injection Sensorless Rotor Position Estimate," 2022 IEEE 17th Conference on Industrial Electronics and Applications (ICIEA), Dec. 2022, pp. 920–925.
56. G. Boztas, M. Yildirim, and O. Aydogmus, "Design and Optimization of a PMSM for Obtaining High-Power Density and High-Speed," TJST, vol. 15, no. 2, Art. no. 2, Sep. 2020.
57. W. Zhao, X. Wang, C. Gerada, H. Zhang, C. Liu, and Y. Wang, "Multi-Physics and Multi-Objective Optimization of a High Speed PMSM for High Performance Applications," IEEE Transactions on Magnetics, vol. 54, no. 11, pp. 1–5, 2018.
58. X. Sun, Z. Shi, G. Lei, Y. Guo, and J. Zhu, "Analysis and Design Optimization of a Permanent Magnet Synchronous Motor for a Campus Patrol Electric Vehicle," IEEE Transactions on Vehicular Technology, vol. 68, no. 11, pp. 10535–10544.
59. A. E. Fitzgerald, C. Kingsley, and S. D. Umans, Electric machinery, 6th ed., International ed., in McGraw-Hill series in electrical and computer engineering. Boston, Mass.: McGraw-Hill, 2009.

Chapter 5
Conclusion

This book has systematically reviewed the rotor pole pattern topology technology for PM machines, presenting a comprehensive framework to address key challenges such as high torque output, minimizing torque ripple, enhancing efficiency, and optimizing structural mass. Through detailed theoretical analyses, numerical modeling, and experimental validation, some unique insights and new methodologies are presented in this book, which aligns with the growing demands of modern industrial applications, such as electric propulsion, robotics, and aerospace engineering.

The research began with an in-depth discussion of the fundamental principles of PM electric machines, emphasizing their advantages in efficiency, reliability, and noise reduction. Central to this study was the classification and optimization of rotor pole patterns, which play a pivotal role in determining machine performance. Several primary configurations are investigated: radial, spoke-type, Halbach arrays and their variants. Building on these typical designs, we proposed some innovative solutions, such as the quasi-cylindrical pole pattern, which effectively reduces torque ripple and cogging torque, while maintaining high average torque output. Similarly, the hybrid-irregular pole pattern was introduced as a solution to balance torque enhancement and ripple reduction by combining multiple magnetization strategies. Additionally, the application of harmonic injection techniques with edge thickness variations was explored, demonstrating improvements in airgap flux density, reduced vibration, and overall performance stability.

Beyond individual pole pattern designs, this study extends its approach to advanced structural optimization techniques. A multi-material topology optimization framework based on variable density representation was presented, which offers a novel approach to optimize the distribution of magnetic and structural materials within the rotor. This method not only enhances torque density but also contributes to the development of lightweight machine structures. Efficiency-focused optimization strategies were also introduced, including the detailed optimization of stator

P. Xiang et al., *Rotor Pole Pattern Topology Technology for Permanent Magnet Electric Machine*, https://doi.org/10.1007/978-981-95-2018-3_5

tooth designs, which achieved a critical balance between torque output and thermal management under rated conditions. Furthermore, the concept of compound Halbach arrays was presented, leveraging enhanced self-shielding effects to reduce rotor mass and improve magnetic flux utilization.

The proposed designs and optimization methodologies were validated through experimental studies, including the development of prototypes and test-rigs. These experiments confirmed the theoretical and numerical predictions, showcasing great improvements in torque density, reduced torque ripple, and increased material utilization, compared to the corresponding conventional designs. The findings underline the practical applicability of these novel designs, and provide a solid foundation for future advancements in PM machines.

As we conclude, it is important to note that the methodologies and insights presented in this book extend beyond the specific designs discussed. The approaches to topology optimization, harmonic analysis, and structural innovation are broadly applicable to various linear and rotary machine configurations, offering pathways for further research and development. Future work could explore the integration of these methods with emerging technologies, such as advanced control strategies and new material systems, to further enhance the capabilities of PM electric machines. In summary, this book provides a robust and innovative contribution to the field of electric machine design. By combining theoretical rigor with practical validation, it offers valuable guidance for researchers, engineers, and practitioners aiming to develop high-performance, efficient, and lightweight PM electric machines for a wide range of modern applications.

Index

P. Xiang et al., *Rotor Pole Pattern Topology Technology for Permanent Magnet Electric Machine*, https://doi.org/10.1007/978-981-95-2018-3

MIX
Papier aus verantwortungsvollen Quellen
Paper from responsible sources
FSC® C105338

If you have any concerns about our products,
you can contact us on
ProductSafety@springernature.com

In case Publisher is established outside the EU,
the EU authorized representative is:
Springer Nature Customer Service Center GmbH
Europaplatz 3, 69115 Heidelberg, Germany

Printed by Libri Plureos GmbH
in Hamburg, Germany